Rita Donaghy's report to the Secretary of State for Work and Pensions

One Death is too Many

Inquiry into the Underlying Causes of Construction Fatal Accidents

Presented to Parliament by the Secretary of State for Work and Pensions by Command of Her Majesty July 2009

Cm 7657 £19.15

ISBN: 978 0 10 176572 5

CONTENTS

This is my independent report. Its contents, including any opinions and/or conclusions expressed, are mine and do not necessarily reflect Government views or policy.

GLOSSARY

ACAI	Association of Consultant Approved Inspectors
BERR	Department for Business, Enterprise and Regulatory Reform which as of June 2009 became part of a new Department for Business, Innovation and Skills (BIS)
BIS	Department for Business, Innovation and Skills (established June 2009)
CBH	Constructing Better Health
CDM	Construction (Design and Management) Regulations 2007
CIS	Construction Industry (tax) Scheme
CITB	Construction Industry Training Board, now known as ConstructionSkills
CMS	Common Minimum Standards (from OGC)
COHME	Construction Occupational Health Management Essentials
CONIAC	Construction Industry Advisory Committee
CPS	Crown Prosecution Service
CSCS	Construction Skills Certification Scheme
DWP	Department for Work and Pensions
EASI	Employment Agency Standards Inspectorate
GLA	Gangmasters Licensing Authority
HEFCE	Higher Education Funding Council for England
HMRC	Her Majesty's Revenue & Customs
HSC	Health and Safety Commission (merged with the HSE in April 2008)
HSE	Health and Safety Executive
HSWA	Health and Safety at Work etc. Act 1974
IoD	Institute of Directors
LABC	Local Authority Building Control (or Building Control)
MEWP	Mobile Elevated Work Platform
NDPB	Non-Departmental Public Body
NVQ	National Vocational Qualification
OGC	Office of Government Commerce
ONS	Office of National Statistics
PPE	Personal Protective Equipment
RIDDOR	Reporting of Injuries, Diseases and Dangerous Occurrences Regulations 1995

SFfC	The Strategic Forum for Construction
SHADs	Safety and Health Awareness Days
SME	Small and Medium-sized Enterprises
THOR GP	Voluntary reporting of occupational diseases by General Practitioners
TUC	Trades Union Congress
UCATT	Union of Construction, Allied Trades and Technicians
VAT	Value Added Tax
WWT	Working Well Together

BACKGROUND

Background

1. I was asked by the then Secretary of State for Work and Pensions the Rt Hon James Purnell[a] to conduct an Inquiry into the Underlying Causes of Construction Fatal Accidents. The Inquiry was announced on 4 December 2008.

2. The Secretary of State referred to "*the unacceptable level of fatalities in the construction industry*" and indicated that, as independent Chair of the Inquiry, I was free to make any recommendations I saw fit. While the focus of the Inquiry should be firmly set on fatalities in construction, stakeholders such as trade unions, industry and campaign groups should be able to contribute pertinent issues at each stage of the work and also be informed of progress.

3. In addition, the Health and Safety Executive was asked to ensure that the resources and expertise of HSE were made available to me throughout my work. The Secretary of State indicated that the HSE "*rightly plays a central role in preventing fatalities in the construction industry*" and that it had already begun the process of reviewing existing information on construction fatalities. It was important that I felt able properly to examine the HSE's work. In addition to the HSE's input, I was able to call upon the expertise of independent academic peer reviewers who would help me carry out the work. A phased approach to the work was outlined and proceeded as follows:

Phase 1

a) A comprehensive review of recent work to consolidate and summarise existing knowledge of causal factors in construction fatal accidents based on HSE internal and externally commissioned work over the last 10 years, including a 2008 study examining the root causes of 25 migrant / foreign worker fatalities.

b) An independent peer review of the Phase 1 report including recommendations for the conduct and scope of Phase 2. The review was undertaken by three academics who provided independence and continuity throughout the work, including direction and oversight of Phase 2.

---

[a] The Rt Hon Yvette Cooper was appointed Secretary of State for Work and Pensions in the Government reshuffle on 9 June 2009 as this report was in the final stages of preparation.

Phase 2

c) Research examining recent fatal accidents in construction and wider sources of evidence:

- A review and analysis of a further '25' recent construction fatal accidents focusing on underlying causes, undertaken by HSE with independent direction and oversight from the Chair and peer reviewers.
- A review of evidence external to HSE from industry / international / insurer / company / trades union sources about root causes of construction accidents and levers within and beyond health and safety systems to make further improvements.

d) Review and oversight of all Phase 2 work by the three independent academic peer reviewers.

Phase 3

e) Chair's review of all aspects of the Phase 1 and 2 work and extensive stakeholder consultation across the industry.

f) Chair's report to DWP Ministers and the HSE Board in June 2009.

Approach to the Work

4. I was very conscious, when considering the approach to the Inquiry, that there was a wealth of experience and a number of reports which had been published on the construction industry. I am aware that the reports by Sir Michael Latham 'Constructing the Team'[b] and Sir John Egan 'Rethinking Construction'[c] have been important influences in the development of the construction industry and I have used their knowledge and wisdom extensively. Latham and Egan both covered the construction industry as a whole, whereas my remit has been a more specific area about the underlying causes of construction fatal accidents. I have not attempted to review the progress, or otherwise, that has taken place in the industry as a whole, in the 15 years since Latham or the two Egan reports of 11 and 7[d] years ago. However, there are some recurring themes which make it clear that there have been major improvements in health and safety in the larger construction companies and some significant improvement at the top end of the industry in welfare facilities for workers and the improved appearance of sites through the Considerate Constructors Scheme. Equally, issues such as

---

[b] Latham, M. 'Constructing the team', Department of the Environment, HMSO, 1994.

[c] Egan, J. 'Rethinking construction', Department of the Environment, Transport and the Regions, HMSO, 1998.

[d] The Strategic Forum for Construction. 'Accelerating change', 2002.

training and skills, pre-qualification, team working and the extent of self-employment and fragmentation in the industry remain key issues which have not been resolved and are still topics which are hotly debated. Some of these issues could well have causal links with fatal accidents in the industry.

5. On the political front the considerable body of knowledge accumulated by the House of Commons Work and Pensions Committee and the Business and Enterprise Committee on the construction industry was invaluable. I would like to acknowledge particularly the work of the then Deputy Prime Minister, The Rt Hon John Prescott, for initiating the Construction Summit in 2001 which led to so many changes.

6. In considering which stakeholders I should meet in the first four months of the Inquiry, I took advice from the HSE and leaders in the industry. In subsequent meetings other organisations and individuals were drawn to my attention. I have made every effort to meet as many people as possible but, in the time available, I was not able to talk to everyone who wished to meet me.

7. The appointment of the independent academic peer reviewers was a matter of urgency. After receiving the CVs of a panel of academics with experience in this field, I decided to appoint three - Professor Andrew Hale, Dr Sonia McKay and Professor David Walters (copy of summarised CVs attached as Appendix 1). The original terms of reference suggested two academics. However, given the amount of work and the time constraints I decided that three would be preferable.

8. The Secretariat of the Inquiry were as follows:

Nicola Walters	Secretary to the Inquiry	HSE
Louise Brearey	Head of Construction Sector	HSE
Indi Patel	Administration Assistant	HSE
Helen Bolt	Consultant supporting HSE's Construction Sector	
John Stevenson	Communications Press Officer	DWP

9. The first formal meeting with the academic peer reviewers took place on 19 January 2009. The first draft of the Phase 1 comprehensive review[e] was available and consideration was given to the appointment of a university to conduct Phase 2 of the Inquiry – external research of the industry covering international, insurer and company evidence. After

[e] Health and Safety Executive. Phase 1 Report: 'Inquiry into the Underlying Causes of Construction Fatal Accidents - A comprehensive review of recent work to consolidate and summarise existing knowledge', 2009. http://www.hse.gov.uk/construction/inquiry.htm

consideration, Loughborough University with a research team led by Professor Alistair Gibb was appointed with a tight timetable to produce the Phase 2 external research[f].

10. It became clear at the first meeting that much of the work would have to be carried out at the same time rather than sequentially in order to try to meet the deadline. Everyone agreed that this was the only feasible approach in the circumstances and, although it meant that the work was more complex, I do not believe it detracted from its quality, or the validity of the final supporting documents.

11. From January to mid-April I spent the major part of my time meeting stakeholders – given the size and diversity of the industry I cannot pretend to have met every group or representative. The people I spoke to on the telephone or in meetings are listed as Appendix 2. Nevertheless, I do believe I have met a cross section of the people who are committed to the construction industry. I have met representatives of Government Departments, Local Government, stakeholders in Scotland and Wales. Although the Inquiry did not actively invite written evidence, nevertheless I did receive some. These contributions are listed in Appendix 3. I have read as much background literature as possible - listed in Appendix 4. I spent three days on site visits with HSE Inspectors, two in London and one in Letchworth.

12. I met trainers, health and safety experts, lawyers and one coroner. I spent one morning at the Old Bailey watching the conclusion of a fatal accident case, which had taken five years to complete. Finally I met six families, and their representatives, who had lost a son or husband in a construction fatal accident.

13. The academic peer reviewers have been very active since their appointment and have worked with Loughborough University and the HSE very closely acting as advisers in overseeing much of the work. The reviews and analysis of the 25 recent construction fatalities (in practice it was 28) turned out to be a considerable task involving interviews with the original inspectors, constructing a suitable model for comparison and testing each case in detail against the model. The model was also checked against the 50 cases already covered in Phase 1 to ensure integrity. I believe the work done represents a considerable contribution to this subject[g]. The result may surprise or disappoint but I

---

[f] Loughborough University. Phase 2 Report: 'Health and Safety in the Construction Industry: Underlying causes of construction fatal accidents – External research', 2009. http://www.hse.gov.uk/construction/inquiry.htm

[g] Health and Safety Executive. Phase 2 Report: 'Inquiry into the Underlying Causes of Construction Fatal Accidents – Review and sample analysis of recent construction fatal accidents', 2009. http://www.hse.gov.uk/construction/inquiry.htm

am convinced it will stand up to scrutiny and should be the subject of continuing work.

14. There are two areas which I have not covered in the Report in detail – international comparisons and insurance implications.

    a) Although the international comparisons elicited by Loughborough University are interesting and thought provoking, I do not believe that existing comparative research is sufficiently comprehensive to learn particular lessons for the UK or for me to make recommendations. One would need to know the context of the whole built environment of a country before valid comparisons could be made which might affect a whole industry.

    b) Secondly, although I had an interesting meeting with the Association of British Insurers about how insurance companies responded to safety in the construction industry, I became conscious that there was insufficient published information available on which to base reliable conclusions or recommendations. I advised the research team at Loughborough University accordingly as it would have meant spending a disproportionate amount of their time on this topic. Insurance is a highly competitive industry with specialist insurance covering large construction projects and low margin premiums covering the bottom end. I do believe it is worthy of further exploration however, if only to explore discount schemes and safety incentives or possible sponsorship of safety advisors.

15. The Work and Pensions Committee, and in particular its Chair Terry Rooney, have taken a close interest in the progress of the Report and invited me to participate in a session at the House of Commons in January. The then Secretary of State, the Rt Hon James Purnell, and the Parliamentary under Secretary of State (Lords), Lord McKenzie, have also taken a close interest in the proceedings and I have kept them informed on a regular basis of my activities and, latterly, my emerging thoughts. As the Department for Business, Enterprise and Regulatory Reform (BERR)[h] also has an important remit in this area I have kept contact with the Minister Rt Hon Patrick McFadden and then Minister Ian Pearson.

---

[h] BERR became part of the Department for Business, Innovation and Skills (BIS) created on 5 June 2009.

EXECUTIVE SUMMARY

16. Everyone in the UK is touched by construction work in one way or another from domestic work through to schools and hospitals and to national or international prestige projects. Everyone has a view about it but there is no sense of shock at the regular toll of fatalities in the industry. We should aim to raise the profile of these tragedies so that a construction fatality becomes socially unacceptable.

17. The work of the Health and Safety Executive is vitally important in ensuring both compliance and culture change and in encouraging leadership and worker involvement in the industry. It must have the resources to continue to do that work, particularly as we emerge from a recession when the number of fatalities tends to rise. However, the HSE cannot succeed in eliminating fatalities without the support of the population as a whole and the Government. This is a social issue and is too important to be confined to the narrower focus of health and safety.

18. Those parts of the industry which are leading the way, both large and small companies, should be supported by ensuring that those out of sight below the Plimsoll line are more extensively monitored and standards improved. To that end I recommend extending the Building Regulations so that health and safety processes should be included when considering building control applications or building warrants.

19. Where construction companies have taken the lead in safety management it is clear that fatalities are reduced. The industry has chosen to adhere to a top-down structure where the Principal Contractor sub-contracts the majority of work on a project. Employment security has made way for maximum flexibility; self-employment, whether genuine or bogus, forms a substantial part of the workforce pattern. Worker representation is weak and the prospect of a bottom-up safety culture being successful in the near future is unlikely. The industry should renew its efforts to establish genuine consultative frameworks to encourage greater worker participation. Only with good leadership and worker participation combined can a safety culture become established. The responsibility for safety already lies clearly with the contractor and this responsibility needs to be further clarified in order to raise standards and assist the courts when considering alleged breaches of health and safety.

20. I recommend that there should be positive duties on directors to ensure good health and safety management through a framework of planning, delivering, monitoring and reviewing. The duties should be explicit so that everyone knows what is expected and breaches can be more clearly identified. As with most advances in society, e.g. seat belts in cars, drink

driving, there comes a time when good practice has to become a legal requirement. I believe the time is right to introduce a clearer sign that society wishes to prevent fatalities and demands a higher standard of behaviour from those in the construction industry who do not at present follow good practice. The major contractors have nothing to fear from such a regulation. Indeed it might help them to raise standards.

21. The courts should be assisted on the issue of specific responsibilities of directors to carry out corporate health and safety governance so that they can make appropriate judgements about the verdict or level of fine or whether disqualification is merited and, if so, for how long.

22. I recommend that the remit of the Gangmasters Licensing Regulations should be extended to include construction. Alternatively, a Regulation should be made which has the same effect. The further down the sub-contracting chain one goes the less secure the worker and the less satisfied with the management of health and safety on site. Society should accept that there needs to be a standard below which no construction worker should have to work. The Gangmasters Licensing Authority would need resources to take on this work and some consideration may have to be given to its existing constitution to ensure it is fit for purpose.

23. Office of Government Commerce guidance and the mandatory Common Minimum Standards should apply throughout publicly funded construction projects, including local authorities, and systems for accountability should be more effectively monitored and enforced with appropriate sanctions. Public procurement is important because of its size and its potential for insisting on driving up standards including health and safety. Those bodies which are monitoring the effectiveness of the mandatory OGC Common Minimum Standards – the OGC itself and the Public Sector Construction Clients' Forum – should be supported in their endeavours. This can only be done by greater accountability of how tax payers' money is spent and effective means of identifying those who do not follow the mandatory standards.

24. Government should not underestimate the impact of the current taxation regime for self-employed workers in construction either on revenue collecting or on other taxpayers. The political will to grasp this particular nettle would be an important step to signal to the industry, and its workers, that society expects standards to be improved and obligations fulfilled.

25. Support should be given to those organisations working on developing the worker safety advisor scheme. The role of Worker Safety Advisors was welcomed by trade unions and many employers. Their future role depends on continued funding but there was no evidence that this would

be available from Government, HSE or employers and trade unions. Dame Carol Black's recent report (Working for a Healthier Tomorrow) included a possible expanded role for trade union safety representatives to promote the benefits of employer investment in health and well-being. I further recommend that any initiative arising from Dame Carol's report should include some funding for a project in the construction industry.

26. The positive role that trade unions can play in health and safety is not fully appreciated by the construction industry and more should be done, particularly by the larger companies, to encourage joint working with the unions.

27. Where there is no trade union presence, renewed efforts should be made to encourage genuine worker involvement so that workers are alert to risk and can speak out, without unfair consequences, about unsafe practices. This would involve the Strategic Forum for Construction, the HSE and any other organisation currently involved in worker engagement.

28. There should be targeted safety campaigns for specific groups of workers who are regarded as vulnerable e.g. migrants, young people, the 55-60 plus age group.

29. Occupational Health remains a serious problem in the construction industry. Thousands of workers die every year from mesothelioma and other occupational cancers and lung diseases. Twenty skilled workers (electricians, plumbers etc) die every week from asbestos related disease and 12 more construction workers die every week from silica related lung cancer. While this was outside the scope of this Inquiry, it is vital that renewed efforts should be made to tackle this issue. The dangers are known and the preventative work needs to be done.

30. The Construction Skills Certification Scheme card system is an important development in the industry and should be supported. However there is a general feeling in the industry that the system needs to be consolidated and renewed with a more strategic approach as to its future development. There are obvious concerns about the fraudulent use of cards and about their application e.g. a basic card could be used by someone undertaking a task requiring considerable skills. However, CSCS represents a commitment by the industry to improve and deserves continuing support.

31. The work of ConstructionSkills is also very important to the industry. It would be useful to see if the allocation of grants could be more redistributive as evidence suggests that it is the larger companies which make most use of the available funding. It would be also useful to know why there is such a high drop-out rate from apprenticeships despite high

volumes of applications. A well trained workforce is likely to be a safer workforce.

32. There should be a review by Higher Education Funding Council for England (and the equivalent bodies in Scotland and Wales), the industry and professional bodies on the adequacy and relevance of university or college curricula on undergraduate and postgraduate construction related courses. The review should look at whether health and safety is appropriately covered in the curriculum, including design and maintenance implications and site awareness not just technical and legal implications. It should also look at whether the Construction (Design and Management) Regulations 2007 are appropriately covered, that assessors are aware of developments in health and safety awareness and that moderators ensure that work is carried out in practice.

33. There should be an investigation into the built-in delays in the system leading to prosecution and conviction or other outcomes on construction fatal accidents. The investigation should include the HSE, coroners, police, Crown Prosecution Service, the legal profession and courts so that there is a thorough picture. The experience that bereaved families have to go through is made worse by their experience of the legal system and recommendations should be made to produce timely and proportionate conclusions.

34. Construction workers, as individuals, should consider the impact on their families of unsafe working practices and should accept responsibility for their own safety so far as they are able. If employees joined a trade union their families are more likely to receive support and advice in the unfortunate event that they suffer a fatal accident; similarly a self-employed worker should have sufficient insurance to enable their families to obtain legal advice should it prove necessary.

35. Construction companies should hold the names and contact numbers and addresses of close family members in the event of accidents.

36. There should be standard agreed bench-marks to test against the myriad of pre-qualification schemes so that sub-contractors do not have to acquire a host of pre-qualifications before participating in tenders for public and private work including local schools, hospitals or housing associations. The time and money spent on satisfying the various pre-qualification schemes divert attention from practical health and safety on the ground. There should also be standard agreed bench-marks for testing health and safety competence in the various schemes. The groups considering common core competencies for pre-qualification should try to co-ordinate their efforts and reach agreement on acceptable transferability of key areas. The Government should take the lead on this as a major client in public procurement. Health and safety,

along with financial stability and capacity, should be an absolute in all common standard requirements.

37. Construction is more than just an industry. It is about the built environment and how we live in it. There should be more political focus on construction and a resource to provide an overview of the various Government Departmental activities. As Government performs the role of client, regulator and provider it would not be practical to make major machinery of Government changes. However, part of ensuring that construction fatalities become socially unacceptable is to raise the profile of construction in Government. This would go a long way to convincing the industry and the population at large that construction is important. There should be a full-time Minister for Construction with a co-ordinating brief to lead on construction. While supporting the general principle of creating the post of a Chief Construction Officer, it is important that there is some clarity and focus to the position. It is doubtful whether one person, no matter how experienced, could fulfil the role of champion, gatekeeper, co-ordinator and Whitehall monitor all at the same time. The important point is that there should be a resource to make the construction role effective and to support the full-time Minister for Construction.

38. It is important that focus remains on developing and refining technological support for the industry. Despite considerable advances in technology in the construction industry over the last ten years, it is clear workplace equipment and machinery are involved in a number of fatal accidents. The HSE Board is considering ways in which to make safety alerts (when specific defects or unsafe circumstances are identified) more effective and disseminated more widely and this work should be supported. The so-called 'legacy' issues, whereby second hand equipment or machinery is still in circulation and could cause further accidents or fatalities, is a significant challenge. More work needs to be done by the HSE, the Strategic Forum for Construction and other appropriate organisations to tackle this 'legacy' challenge. Individuals or companies hiring this equipment should also ensure that they are fully aware of the capabilities of such machinery and its possible dangers and that they are sufficiently trained to identify any faults in the equipment. I recommend that further work should be undertaken in the areas outlined in Paragraph 209.

39. The industry should continue to support partnership working through the supply chain. Working Well Together is a good example of the industry and HSE working together to spread good practice particularly in the use of workplace equipment.

40. It is a disgrace that we have such a low level of reporting serious accidents, let alone near-misses, and is indicative, not just of ignorance and commercial factors for both the worker and the company, but of society's attitudes to workers' injuries. If we had a higher proportion of reporting serious accidents, it might help us to achieve a more accurate picture about fatalities. I recommend an awareness raising campaign so that individual workers and companies take the issue of reporting accidents more seriously.

41. The issue of resources for the HSE is a political one. However a number of issues have been summarised in Paragraph 227 which, if proceeded with, would require additional resources. The independence and professionalism of the HSE inspectors is important and should be retained. There are insufficient resources in London to carry out even the existing workload; this has been a problem for some years. The HSE should take steps to improve this situation.

42. There should be a pilot study by the HSE to determine the impact of more non-accident prosecutions.

43. There should be a review in the HSE of the communications strategy about fatal accidents to ensure a more co-ordinated and easily absorbable presentation. More opportunities could be found to publicise outcomes of legal cases and 'hot-spot' accidents.

44. HSE should encourage its inspectors to promote the principles of the joint Institute of Directors / Health and Safety Commission guidance for directors and leaders of companies by the use of more easily absorbable presentation material. Awareness of the guidance should be raised in the short-term pending the positive legal duties on directors.

45. The review of recent construction fatal accidents represents an important contribution to this Inquiry. Work should continue in this area on a regular basis as it will further illuminate the underlying causes of fatalities. The most frequently cited in the study are the incidence of training factors, experience, information and advice deficiencies, risk perception, rescheduling of work without planning, minor / one-off jobs, compliance, equipment operability, space, personal protective equipment (PPE) issues and tools not designed to fit the user / task. Most of these accidents were preventable.

RECOMMENDATIONS

1. I recommend extending the Building Regulations so that health and safety processes should be included when considering building control applications or building warrants.

2. I recommend that the remit of the Gangmasters Licensing Regulations should be extended to include construction. Alternatively, a Regulation should be made which has the same effect.

3. I recommend that there should be positive duties on directors to ensure good health and safety management through a framework of planning, delivering, monitoring and reviewing.

4. I recommend that the courts should be assisted on the issue of specific responsibilities of directors to carry out corporate health and safety governance so that they can make appropriate judgements about the verdict or level of fine or whether disqualification is merited and, if so, for how long.

5. I recommend there should be an investigation into the built-in delays in the system leading to prosecution and conviction or other outcomes on construction fatal accidents.

6. I recommend that there should be a full-time Minister for Construction with a co-ordinating brief to lead on construction.

7. I recommend that Office of Government Commerce guidance and the mandatory Common Minimum Standards should be applied throughout publicly funded construction projects, including local authorities, and systems for accountability should be more effectively monitored and enforced with appropriate sanctions.

8. I recommend that there should be standard agreed bench-marks to test against the myriad of pre-qualification schemes so that sub-contractors do not have to acquire a host of pre-qualifications before participating in tenders for public and private work including local schools, hospitals or housing associations. The Government should take the lead on this as a major client in public procurement.

9. The work of ConstructionSkills is also very important to the industry. I recommend that there should be a review of the allocation of grants as evidence suggests that it is the larger companies which make most use of the available funding. Some more redistributive allocation would be welcome.

10. I recommend that some research is carried out to identify why there is such a high drop-out rate from apprenticeships despite high volumes of applications.

11. The Construction Skills Certification Scheme card system is an important development in the industry. I recommend that the system should be further consolidated and renewed with a more strategic approach as to its future development.

12. I recommend there should be a review by the Higher Education Funding Council for England (and the equivalent bodies in Scotland and Wales), the industry and professional bodies on the adequacy and relevance of university or college curricula in covering design, health and safety awareness and risk management issues.

13. I recommend that the industry should renew its efforts to establish genuine consultative frameworks to encourage greater worker participation.

14. I recommend that support should be given to those organisations working on developing the worker safety advisor scheme. Dame Carol Black's recent report (Working for a Healthier Tomorrow) included a possible expanded role for trade union safety representatives to promote the benefits of employer investment in health and well-being. I further recommend that any initiative arising from Dame Carol's report should include some funding for a project in the construction industry.

15. The positive role that trade unions can play in health and safety is not fully appreciated by the construction industry and I recommend that more should be done, particularly by the larger companies, to encourage joint working with the unions.

16. Where there is no trade union presence, I recommend that renewed efforts should be made to encourage genuine worker involvement so that workers are alert to risk and can speak out, without unfair consequences, about unsafe practices. This would involve the Strategic Forum for Construction, the Health and Safety Executive and any other organisation currently involved in worker engagement.

17. I recommend that construction workers, as individuals, should consider the impact on their families of unsafe working practices and should accept responsibility for their own safety so far as they are able. I recommend that employees should join a trade union as their families are more likely to receive support and advice in the

event of a fatal accident; similarly a self-employed worker should have sufficient insurance to enable their families to obtain legal advice should it prove necessary.

18. I recommend construction companies should hold the names and contact numbers and addresses of close family members in the event of accidents.

19. I recommend that the industry should continue to support partnership working through the supply chain.

20. While this was outside the scope of this Inquiry, Occupational Health remains a serious problem in the construction industry with thousands of workers dying every year. I recommend that renewed efforts should be made to tackle this issue.

21. I recommend that more work needs to be done by the HSE, the Strategic Forum for Construction and other appropriate organisations to tackle the 'legacy' challenge of second hand equipment or machinery still in circulation. Individuals or companies hiring this equipment should also ensure that they are fully aware of the capabilities of such machinery and its possible dangers and that they are sufficiently trained to identify any faults in the equipment.

22. I recommend an awareness raising campaign so that individual workers and companies take the issue of reporting accidents more seriously.

23. I recommend that there should be targeted safety campaigns for specific groups of workers who are regarded as vulnerable e.g. migrants, young people, the 55-60 plus age group.

24. There are insufficient HSE resources in London to carry out even the existing workload; this has been a problem for some years. I recommend that HSE should take steps to improve this situation.

25. I recommend that there should be a pilot study by the HSE to determine the impact of more non-accident prosecutions.

26. I recommend that there should be a review in the HSE of the communications strategy about fatal accidents to ensure a more co-ordinated and easily absorbable presentation.

27. I recommend that HSE should encourage its inspectors to promote the principles of the joint Institute of Directors / Health and Safety Commission guidance for directors and leaders of companies by the use of more easily absorbable presentation material.

28. The review of recent construction fatal accidents represents an important contribution to this Inquiry. I recommend that work should continue in this area on a regular basis as it will further illuminate the underlying causes of fatalities.

GENERAL INTRODUCTION

46. The Construction Industry generally is modelled to provide maximum flexibility. Consequently the majority of functions are contracted out and at least 40% of workers are self-employed or CISs[i]. The advantages are obvious in that it reduces overheads. Some but not all argue that it improves profitability and productivity. The disadvantages are that it becomes more difficult for a safety culture to flourish, worker engagement is weak, employment security and continuity is minimal and skills training is at best patchy. There are many organisations and pieces of research arguing that recruitment to the industry is too narrow and would benefit from more women, ethnic minorities and people with certain disabilities. While, in my view, this is undoubtedly true, the current model of the construction industry would make substantial progress in this area very difficult[j].

47. There have been some important incremental improvements in safety as the ten year review[e] outlines. The Latham[b] and Egan[c,d] reports led to step changes in the industry, and political initiatives taken such as the Prescott summit were significant.

48. Many improvements have been made in safety management and compliance in the construction industry and the reduction in the number of fatal accidents since the early 1990s is an important achievement. However, we should not continue to rely entirely on safety management to bear down on the number of fatalities each year. This will never eliminate fatalities or reduce significantly the number of serious accidents. There is an important distinction between safety management and safety leadership. Leadership and culture change are essential if we are to eliminate fatal accidents and the construction industry will need some support from the Government and wider society to achieve this. Safety leadership at local and national levels will help to change public attitudes.

---

[i] The Construction Industry Scheme (or CIS for short) is a tax scheme that affects contractors and sub-contractors in the construction industry who carry out construction operations. Under the scheme sub-contractors must register with the Inland Revenue and obtain either a registration card or a certificate and show this to the contractor before they can be paid for the work they have done. The contractor will then deduct an amount on account of tax and Class 4 National Insurance, and pay the net amount (extracted from http://www.hmrc.gov.uk/pdfs/hmc_const.pdf - see also Inland Revenue leaflets IR56, entitled 'Employed or Self-Employed' and IR148 entitled 'Are your workers employed or self-employed? A guide for tax and National Insurance for contractors in the construction industry')

[j] Clarke, L. and Gribling, M. 'Obstacles to diversity in construction: the example of Heathrow Terminal 5', *Construction Management and Economics*, July 2008.

49. The UK Contractors Group[k] and some others have accepted responsibility for driving a safety culture onto their sites by improved site management training, encouraging behavioural change and by adopting a policy of requiring some health and safety induction and skills before entry to their sites. This top-down approach to safety management relies very heavily on the appropriate number and quality of site managers, motivated by drive from the top of their organisations and through the supply chain. It is accepted in the industry that site management training is a top priority and there is little doubt that this good work will continue through the UK Contractors Group. However there is a shortage of good site managers and this will remain a vital weakness in the system for some time to come.

50. Another feature of the construction industry, with a few exceptions, is the absence of pre-planning and integrated teams before work starts on site – points constantly raised by Latham[b] and Egan[c]. One stakeholder referred to the industry as having "*an incredibly slow clock speed. One learns from experience on projects but a site manager might do 30-40 projects in a working lifetime. If they are large projects it might be only 10-20 projects in a working life*". The stakeholder added "*The industry consists of a lot of practical people doing things – problem solving rather than problem anticipating. We are not good at forward thinking*". This situation underlines the importance of the Construction (Design and Management) Regulations 2007[l] and of appropriate university and college training on design and management, including health and safety, which I refer to in later chapters.

51. I maintain that construction fatalities are regarded as socially acceptable at present for the following reasons:

    a) A construction fatality rarely appears in a national or even regional newspaper when it occurs.

    b) Prosecutions and sentencing are ludicrously low.

    c) Families who suffer a construction bereavement and come up against 'the system' for the first time find themselves virtually unsupported.

    d) Public reaction to the number of fatalities in construction is to say "*it is a dangerous industry*" or "*I'm surprised it* [the number of fatalities] *is as low as that*" or "*there's no entry requirements so they don't get the best people*" – the implication being that construction deaths are not important.

---

k A formal group of major contractors. http://www.ukcg.org.uk

l The Construction (Design and Management) Regulations 2007, SI 2007/320

52. There is a choice to make a change. Construction fatalities should become socially unacceptable and a change in attitude in society should be led by Government. Those already working hard to improve safety in construction – the Strategic Forum for Construction (SFfC), the TUC and trade unions, the safety experts, the HSE and campaign groups – deserve support in this endeavour. Whatever the differences in this fragmented industry – and there are many – this one element should unite all groups - *To prevent death in the construction industry.*

Rita Donaghy CBE FRSA

EXISTING HEALTH AND SAFETY REGULATIONS AND GUIDANCE RELATING TO THE CONSTRUCTION INDUSTRY

53. The key health and safety Regulations relating to the construction industry (inter alia) are listed as Appendix 5. They represent (taken with the guidance that accompanies them) a considerable body of standard setting which, if followed from top to bottom of the industry, would revolutionise safety and eliminate poor standards in construction.

54. The Construction (Design and Management) Regulations 2007 are particularly relevant to the planning and management of health and safety, from concept design through to demolition. CDM 2007 aims to integrate health and safety into the management of all construction projects. It places specific duties on those who can influence health and safety; client, designers and contractors. This approach to risk communication and management throughout the supply chain is their distinctive feature. The principal focus is to identify and eliminate hazards and effectively manage residual risk. It requires that those appointed by the client should be competent and encourages co-operation and co-ordination between duty holders. The regulations advocate a proportionate approach based on degree of complexity and risk, discouraging unnecessary bureaucracy. For larger projects there is the requirement to appoint a CDM Co-ordinator and Principal Contractor and notify the enforcing authority of the project. Specific duties for on-site health and safety risk during the construction phases are included at part 4 of CDM 2007.

55. Another area of considerable importance is the guidelines issued by the Office of Government Commerce (OGC) which is part of HM Treasury. The OGC is responsible for improving value for money by driving up standards and capacity in public procurement. Guidance note 10 in the OGC Construction Procurement Guidance - Achieving Excellence series[m] identifies the planning, management and review requirements that Government Construction Clients must adopt which requires:

   a) All projects to be in accordance with OGC Achieving Excellence Guidance.

   b) Collaborative relationships – traditional contracts should seldom be used and only where it offers best value.

   c) Pre-qualification of contractors.

   d) Monitoring of health and safety performance.

   e) Review.

---

[m] Office of Government Commerce. 'Achieving excellence in construction procurement guide: Health & safety', AE10, 2007.

f) Use of expert support.

g) Collection of health and safety performance data from all parts of the supply chain.

h) Requiring Construction Skills Certification Scheme (CSCS) or similar means of demonstrating competence.

56. It is made mandatory by the 'Common Minimum Standards for Procurement of the Built Environment' (CMS)[n], issued by the OGC. They apply whether the project is funded through capital procurement, a private developer scheme or a Public Private Partnership / Private Finance Initiative.

57. The Common Minimum Standards do not apply in the devolved authorities of Scotland and Wales, nor do they apply to projects carried out overseas. In Scotland, the Scottish Government has produced the Construction Procurement Manual which applies mandatory policy and procedures to Scottish Government Directorates and other directly funded public bodies in Scotland (but not including local authorities). This manual contains guidance on health and safety in procurement. The Welsh Assembly is currently preparing its own guidance on public procurement good practice specific to construction.

58. Although the OGC guidance and the mandatory Common Minimum Standards are generally regarded as excellent, the OGC does not have enforcement powers and the quality of reporting varies from Government Department to Department. As the extent of public procurement is significant varying from 30% to 40% of total build, comprehensive information about whether the guidance is followed, particularly on health and safety, is crucial.

---

[n] Office of Government Commerce. 'Common minimum standards for procurement of the built environments in the public sector'. http://www.ogc.gov.uk/construction_procurement_common_minimum_standards_for_the_built_environment.asp

BUILDING CONTROL

59. Many stakeholders made the point that, while there were obvious improvements in construction safety in the larger companies, the main problem was in the small building or refurbishment sector. While the statistics show that this is only partially true, there is no doubt that large areas of the construction industry are out of sight below the Plimsoll line. Those companies which are striving for excellence or even to improve feel let down by the standards of those who are not in trade federations or professional groupings and they want to create a more level playing field. Some have urged discounts on VAT and other incentives to undermine the informal economy. I do not feel qualified to comment on this but I do accept that some more local oversight will be an important step towards improving safety standards in the hard to reach sector. It will also help to raise the consciousness of the public to a new safety culture.

60. I propose that there should be a Building Regulation or an amendment to an existing Regulation, imposing a duty of care on persons carrying out work to do so safely. This Regulation would be enforced by Building Control Surveyors (or Officers). This would extend their enforcement from the safety of what is built to include the safety of the building process.

61. This would mean extending the remit of Building Control to include health and safety requirements and integrating them into the building process. This would not mean that Building Control Surveyors (also known as Building Inspectors or Building Control Officers) would take on the role of a Health and Safety Inspector. The health and safety requirements would be focussed on areas of high risk such as working from heights or shoring up trenches. Further examples are given in Paragraph 66. If the process was included in the Building Regulations, the role of the Building Control Surveyor would be to check that the safeguards were included in the particulars of how a client intended to build and, if they were not, to advise accordingly.

62. This would apply to England and Wales as the situation in Scotland is covered by different Regulations, but arrangements with comparable effect should be made there.

63. I am aware that the HSE, Local Authority Building Control (LABC) and the Association of Consultant Approved Inspectors (ACAI) representing private sector consultants are drawing up a joint working protocol to further develop information sharing aimed to improve standards on site and reduce accidents. These developments are very welcome and build on existing informal relationships in some local areas.

64. There are over 380 different local authorities with different approaches. Compliance policies for Building Control are pro-active with an emphasis on prevention rather than enforcement or prosecution. The private sector Approved Inspectors do not have enforcement powers and have to refer any matters requiring those powers to the local authorities for action. The competitive environment shared by the two groups minimises the serving of notices or prosecutions.

65. Building Control Surveyors inspect construction projects in progress and will advise the client or contractor of any contravention so that it can be rectified. This is done by persuasion and helpful advice and follow up inspections. A formal notice is not usually issued unless the owner or contractor had not acted upon the advice or had covered up the contravention. A similar approach is adopted to work which had not been subject to a Building Regulation application.

66. All grades of Building Control staff inspect construction projects and vet plans within their respective grades. Enforcement policies would usually require fully qualified staff to supervise or carry out that aspect of the work themselves. Building Control staff receive some health and safety training to support them in their work on construction sites but they would not be health and safety specialists. However, qualified staff would be able to identify basic site safety issues regarding working at height, working in trenches or confined spaces, on ladders, or tower scaffolds or on fragile roofs – where most fatal accidents take place.

67. In England and Wales there are around 4,000 Building Control staff, most of whom are Building Control Surveyors visiting construction sites at several different stages of the work. They have the potential for much greater penetration into the small scale or refurbishment market where many projects never come to the attention of the HSE if they are not formally notifiable under CDM 2007.

68. The Department for Communities and Local Government issued a consultation document 'Future of Building Control'[o] in 2008 and work is continuing on this. The emphasis on sustainability is understandable but safer construction sites and safe maintenance of completed buildings is no less important and there is an opportunity here to make a step change in construction safety.

69. There is concern that formally adding the process function to the Building Control Surveyors' responsibilities will cause overload and might even push some clients into the informal economy to get their

[o] Department for Communities and Local Government. 'The future of Building Control: Consultation paper', 2008. http://www.communities.gov.uk/publications/planningandbuilding/futurebuildingcontrol

work done. My view is that if we do not raise our game as a society in our attitude to safety on construction sites the importance of the construction industry will never be fully appreciated and people will continue to die.

70. There would need to be an investment in further training in health and safety on construction sites for Building Control Surveyors to ensure consistency. As this additional power would be very much with the grain of the Building Control Surveyors' work, additional costs should be kept to a minimum and would have to be met by the client.

Scotland

71. The system in Scotland is quite different. Clients and contractors must have a warrant before they can start work. If a client varied from their pre-site warrants without an authorised amendment they could be open to prosecution. The warrant includes the health and safety of the finished product but not the process as this is not a devolved function to the Scottish administration. There are approximately 450 Building Standards Officers in Scotland, all of them public sector employees and they are mainly office based because of the difference in the system. As the process of building is a matter for Westminster, some appropriate consideration will have to be given to any change in authority for Building Standards Officers.

72. It is possible that future changes to Building Control in England and Wales will adopt a similar pattern to that in Scotland i.e. pre-site warrants. This will not conflict with the proposal to add building safety to the powers of Building Control Surveyors. Indeed CDM 2007 encourage pre-planning and preparation including health and safety.

DIRECTORS' RESPONSIBILITIES

73. There are many good examples of construction companies where boards of directors have placed a high priority on health and safety and consider it as one of the first, if not the first, items on their Board agenda. Although even those companies do experience fatal accidents, the overall results of their taking that responsibility for health and safety can be seen in the improved results on serious accidents and on fatalities. Companies will argue that strong leadership on health and safety also makes good business sense and various projects, both public and private, are quoted as delivering on time and under budget with no fatalities and few accidents.

74. However, because of the structure of the industry and the nature of contracting out various functions, many sub-contractors have indicated that they carry the burden of financial and health and safety risk even if the Principal Contractor still has legal responsibility for the project.

75. While major contractors maintain that they are setting an example through leadership, improved site management and by driving a health and safety culture through the industry, sub-contractors claim that this system simply passes the overall risk down the chain. Research on the role of supply chains in improving health and safety management suggests that where such 'direct effects' are sought they are seldom operational beyond first level suppliers and there is little evidence of them being passed on to second or third supply levels, indeed there is more evidence of negative effects of price and delivery conditions being passed down to these levels[p], the implication being that the further down the chain you go the more compromised the financial and safety considerations. The complications arising from this structure is reflected in the legal system for determining any contractual disagreements between contractors and sub-contractors. It is a system which is sophisticated and costly and frequently used – a complete contrast to the legal system for handling construction fatalities.

76. This means that the issue of who is responsible for preventing a construction fatality is a lot less obvious than might at first appear. The temptation for the courts can be to attribute a greater level of personal responsibility onto the deceased construction worker because their role is often clearer than the obfuscation of the role of the Principal Contractor and the myriad companies on site. In smaller companies there is a clear line of sight between directors and the activities on site so it is easier to prosecute a particular director. This point was made to

---

[p] For a recent review of this evidence more generally see, Walters, D. and James, P. 'Understanding the role of supply chains in influencing health and safety at work', Draft report for the Institution of Occupational Safety and Health, Wigston, Leicester – in press.

me several times by smaller companies and their federations. They were not claiming that they were unfairly treated so much as that they were easier targets. On the corporate side the point was put to me by one lawyer that Principal Contractors were always targeted and sub-contractors were let off the hook when prosecutions occurred. Whatever the truth of these two claims - and it is possible that some work could be done to establish this - the precise responsibilities of directors of construction companies for safety are not always appreciated by the courts. Even when the social unacceptability of a construction death is appreciated, it is not always reflected in the fines or number of disqualifications of directors.

77. The issue of directors' responsibilities has been the subject of heated debate for decades and the arguments are well rehearsed on both sides.

a) Are the implied duties on directors contained in various Acts including the Companies Act[q] and Health and Safety at Work Act[r] (HSWA) sufficient to persuade a director of a company to take health and safety seriously?

b) Would the imposition of individual director duties undermine the principle contained in the Health and Safety at Work Act that employers (collectively) have responsibility for the health and safety of their employees?

c) Is not Section 37 of the Health and Safety at Work Act sufficient to prosecute an individual director in certain circumstances?

d) Will the introduction of the Health and Safety (Offences) Act 2008[s] (increasing fines and the risk of imprisonment) not have the desired effect?

78. The current position is that the Health and Safety at Work Act and its accompanying Regulations do not place explicit, legal duties on directors. HSE Inspectors do have the power to prosecute individual directors under the Section 37 of the Act (directors' responsibility for offences committed by their organisations). One of the arguments used against having a legal duty on directors is that there is already provision to disqualify a director found guilty under Section 37 by reason of neglect, consent or connivance. However, for the years 2002-03 to 2007-08 there was a total of 6146 convictions from HSE enforcement in Britain as a whole and of those 61 were guilty verdicts under Section 37[t].

---

[q] Companies Act 2006

[r] Health and Safety at Work etc. Act 1974

[s] Health and Safety (Offences) Act 2008

[t] The total prosecutions for <u>all</u> industries taken under Section 37 for the period 2002-03 to 2007-08 was 96 of which 61 resulted in guilty verdicts, 10 not guilty, 9 adjourned / not concluded and 16 withdrawn / not taken. The breakdown for construction prosecutions for this period is not available at the time of going to press.

The courts do have powers to disqualify directors found guilty of an offence but appear rarely to use this power. In one fatality case a magistrate refused to disqualify a director on the grounds that "*It wasn't as if he had his hand in the till*".

79. The Company Directors Disqualification Act 1986[u], Section 2(1), empowers the court to disqualify an individual convicted of an offence in connection with the management of a company. This includes health and safety offences. This power is exercised at the discretion of the court and is rarely used on health and safety offences. In my view that is because the courts are very reluctant to use this power due to the low status of a construction fatal accident and possibly to a subconscious antagonism to 'elf n' safety'. Similarly Section 37 has not proved an easy part of the Health and Safety at Work Act to successfully prosecute. Even with a recent House of Lords judgement and fresh guidelines to HSE Inspectors, I think it unlikely to increase the number of prosecutions of directors substantially, because of the structure of the industry and the difficulty of identifying responsibility.

80. The Health and Safety Commission considered the effectiveness of the current arrangements in May 2006 and decided it could not, at that stage, recommend to Ministers new legal duties on directors, due to the uncertainty of other developments including discussions on corporate manslaughter, and their potential impact on directors' behaviour. The HSC produced new guidance[v] for directors and board members jointly with the Institute of Directors and this was launched in October 2007. The HSE also issued revised guidance to inspectors reinforcing existing guidance on their ability to enforce against directors and to remind courts of their power to disqualify convicted directors.

81. The HSE Board also agreed to return to the issue of legal duties in due course and its Chair gave a commitment to the Work and Pensions Committee in 2008 that the HSE would revisit the question of director duties "*following a timely, thorough and independent evaluation of the impact of the voluntary approach………. The findings of the evaluation and other relevant evidence offered by stakeholders will be made available to HSE's Board in early 2010*".

82. The Corporate Manslaughter Act[w] and the Health and Safety (Offences) Act are now on the statute book. I believe their impact will be adversely affected by two missing links: firstly the courts will reflect the general view of society that construction deaths are a regrettable fact of life

---

[u] Company Directors Disqualification Act 1986

[v] Institute of Directors and the Health & Safety Commission. 'Leading health and safety at work: Leadership actions for directors and boards members'. HSE Publication INDG417, October 2007.

[w] Corporate Manslaughter and Corporate Homicide Act 2007

rather than preventable tragedies; secondly the precise responsibility of a director to ensure his/her company adopts an effective health and safety structure is not spelled out in a legal duty. The attitude of society may take some time to change but I believe it is imperative that leaders in Government and the media should try to ensure that this happens. The specific responsibilities of directors to carry out corporate health and safety governance already exist in various guidelines but are not legally enforceable. The courts should be assisted on the issue of specific responsibilities of directors to carry out corporate health and safety governance so that they can make appropriate judgements about the verdict or the level of fine or whether disqualification is merited and, if so, for how long.

83. The IoD/HSC guidance, referred to in Paragraph 80 above, was described as "*leadership actions for directors and board members*". It provides a basic framework for action which includes planning, delivering, monitoring and reviewing a board's health and safety management. It advises that "*Board members need to establish a health and safety policy that is much more than a document – it should be an integral part of your organisation's culture, of its values and performance standards*". This framework could form a basis for outlining positive legal duties on directors to promote health and safety as part of corporate governance and to ensure that good policy intentions are turned into explicit changes in organisation, behaviour and technology.

84. One health and safety manager said to me "*Good health and safety governance is part of good corporate social responsibility. Health and safety should be part of the Annual Report and a component of risk management. There might also be a moral equivalent for the environment as well – reasonable arrangements to inform yourself, monitor performance against standards, taking action if you fall below standards.......... I would not do it with fines and imprisonment. Improvement notice, awareness training and the third stage would be dismissal. There would be explicit and published levels of action to improve*".

85. There are options which could be followed to create positive legal duties on directors:

a) A change to the Health and Safety at Work Act which would impose a general duty on individual directors "*to take all reasonable steps to ensure health and safety*";

b) A self-standing Regulation which would outline the framework for health and safety management and risk assessment of the kind outlined in the IoD/HSC Guidelines – 'Leading health and safety at work'.

86. In considering sentencing guidelines for new and existing legislation, courts need to have clear and specific criteria about what directors are expected to do. These responsibilities will assist the courts in coming to a verdict and if appropriate applying penalties. In 2007/08, the Sentencing Advisory Panel consulted widely on the approach to sentencing for the new offence of corporate manslaughter and for health and safety offences resulting in death. The Sentencing Guidelines Council is currently considering the advice of the Panel and is expected to consult on a draft guideline later in 2009. I hope that the opportunity presented for clarifying director responsibilities for health and safety will be taken.

87. Gross negligence manslaughter also applies to individual directors. When individual officers of a company (directors or business owners) by their own grossly negligent behaviour cause death, gross negligence manslaughter can be proved.

88. I believe that the gap in the law is about the need to clarify for the courts what is expected of directors and board members – that they have duties to ensure a framework of good health and safety management in the same way they have fiduciary responsibilities. This would mean that the directors of a Principal Contractor would need to satisfy themselves that health and safety management is adequate in all the companies they use as sub-contractors. It would not mean that they could pass down the risk for health and safety management to their sub-contractors. However it would mean that the directors of the Principal Contractors and sub-contractors must take strategic ownership of their company's health and safety management systems.

WORKERS

Employment Status

89. The employment status and security of construction workers is an important area for this Inquiry. Many stakeholders, particularly trade unions, some academics and bereaved families, feel strongly that self-employment, whether genuine or bogus, adds to the risk in the industry because self-employment is such a high proportion of the total. In London it is approaching 90%. The self-employed can never be genuine apprentices nor could they take on the role of safety representatives. Some claim that the under-reporting of serious accidents is also because the self-employed tend not to report them as they do not receive benefits. The cost of any permanent injury to a self-employed person is probably met by the state. Some claimed that they were less likely to report unsafe practices because "*they wanted a job next Monday*"; in other words they were less secure in their employment. Others, particularly the trade union UCATT, maintained that bogus self-employment in the shape of CIS was a tax evasion system where the Exchequer lost millions, if not billions, and exploitation and unsafe practices were commonplace. Although Latham's report[b] was not about safety or the welfare of workers in the construction industry, he did refer to the extent of self-employment. While appreciating the concerns which had been expressed to his review team he did not feel that he could recommend "*a statutory return to employed status as a general rule*". He felt that those who wished to work on a self-employed basis should be permitted to do so but they must accept the risks which real self-employment involved.

90. He went on to say "*Treasury Ministers have already announced proposals intended to improve the effectiveness of the Tax Deduction Scheme, though it will take four years to introduce the new arrangements. If the effect is to bring about a return to directly employed status by those who were never really self-employed in the first place, that alone may help to improve training arrangements and closer supervision of performance on site. It may also reduce unfair competition for firms which discharge their statutory responsibilities for paying taxes and other charges*".

91. Egan did not refer to direct or self-employment in his Report[c] but he underlined the importance of valuing the people who work in construction if culture change in the industry was to be delivered. He stated "*Not only is the quality of the workforce fundamental to the process of change in construction but also the way workers are treated. In our view, the workforce is undervalued, under-resourced and frequently treated as a commodity rather than the industry's single most*

important asset". He continued "*Clients and their customers do not like the poor image of the industry in this respect* (referring to the *'typically appalling'* facilities available to workers on site) *any more than does the industry itself. It does not require a big step to provide workers with uniforms, proper facilities and rest areas. Construction sites themselves should become advertisements for the industry and the firms working on them*".

92. The Business and Enterprise Committee looked in more detail at the employment issues in 'Construction Matters'[x] (July 2008). It stated "*The status of self-employment defines the relationship between a person and the company they are undertaking work for as subject to commercial rather than employment law. For the individuals concerned, the main motivation is essentially about tax, whereas for contractors engaging self-employed workers, it provides greater flexibility in terms of engagement and contract termination*". The Committee considered the advantages and disadvantages of self-employment (paras 164-166) and referred to the role of Government to provide a steadier stream of work for the industry which could create incentives to take on more direct employees. They also referred to the fact that some firms saw a competitive advantage in direct employment both for the employee and the quality of the end product. However the committee did not come to a definitive view as to whether Government or the construction industry should make any radical changes.

93. It was more robust on the issue of "bogus" self-employment. It described the financial incentives for the industry and the "bogus" self-employed and the fact that there were huge differences in the estimates of how many workers were involved.

94. There seems to be more agreement that "*the problem is particularly acute in the South and London where self-employment constitutes 89% of firms and migrants form 42% of the workforce*".

95. The Committee reported that in April 2007 the HMRC, the Department responsible for the Construction Industry (tax) Scheme (CIS), introduced a "*radical overhaul of the scheme*" where sub-contractors are now required to register on-line rather than individuals carrying CIS cards to verify their registration with the scheme. "*Contractors must now submit a monthly return detailing all their sub-contractors paid during the tax month, and certifying that none of them are in fact employees*". In the view of the Committee, the questions asked of a contractor to establish whether any of their sub-contractors are self-employed, are remarkably similar to the criteria used for identifying direct employment. Views on

---

[x] House of Commons Business and Enterprise Committee. 'Construction Matters', Ninth Report of Session 2007-08, Volume 1 HC127-1, July 2008.

the new system vary and the Committee concludes "*the success of this new approach will depend on the collective 'buy-in' of contractors. Government must also ensure HMRC has the power and resources to monitor and enforce compliance*".

96. The idea that all contractors in the construction industry would somehow act collectively on this one issue when there is little evidence that they act collectively on more fundamental structural issues is, in my view, wishful thinking and it relies too much on HMRC monitoring and enforcement resources which are likely to come under pressure in any economic down-turn. Successive Governments have tried to deal with the issue of "bogus" self-employment, with very little measurable success. It may be that successive Governments see the various schemes they have adopted as a buttress against the huge informal economy in construction - a compromise so that at least some tax is collected. Unfortunately it also conveys a tolerance of the disengagement of the construction company and worker from their tax and insurance obligations. Governments should not underestimate the impact this has on other taxpayers.

97. Granted it is difficult to prove that "bogus" self-employment, of itself, is an underlying cause of fatalities. However it is certainly a factor in low levels of training, job security, the likelihood of reporting serious accidents or unsafe practices or of encouraging team-working in the industry and all these factors are linked to the underlying causes.

98. This is an issue which successive Governments have been unwilling to grasp effectively. If the political will existed and enforcement mechanisms were properly resourced it is probably the single most important step which could be taken to signal to the industry, and its workers, that society expects standards to be improved and obligations fulfilled.

99. Turning to the issue of direct employment in the industry I believe the industry is too diverse to make a blanket recommendation. Stakeholders representing smaller and medium contractors have indicated that they could not afford to employ all of their workers directly where they could not guarantee continuous work as the overheads would be too expensive. Similarly specialist sub-contractors were comfortable with self-employment as their priority was core competence and qualification for the job rather than employment status. To recommend that only major contractors or those applying for public sector contracts should employ directly would seem to me to be too simplistic. Granted some companies see direct employment as giving them a competitive advantage or a method of attracting a higher quality workforce but others do not.

100. Self-employment is a feature across the construction industry. Some only become a self-employed worker because it is the only way they can get a job and they are employees to all intents and purposes. Others choose it as their preferred way of working. In 2007-08 the self-employed proportion of the total number of workers in construction was just over 40%[y]. Personally I believe that it is a conservative figure as it is impossible to give accurate numbers for those moving in and out of the informal economy. This situation is unlikely to change in the near future. Since the Health and Safety at Work Act in 1974[r], there has been such an increase in sub-contracting and outsourcing that I would argue that the section on responsibility for third parties is out of date.

101. Whether or not responsibility for paying tax is separated from responsibility for safety and health, the moral as well as legal obligations for safety rest firmly with the Principal Contractor or contractors using self-employed on their sites. This safety obligation should be the same as if the worker was directly employed. A clear legal outline for directors and leaders of what is expected of them to deliver effective health and safety standards is therefore essential. Without this there is unlikely to be any improvement in the structural conditions of the industry that are a major factor in determining its poor health and safety outcomes.

Safety Reps / Advisors

102. Under the Information and Consultation of Employees Regulations 2004[z], where there are 50 or more employees, an employer is required to inform and consult its workforce on a number of issues if 10% of the workforce trigger negotiations to that effect. The workforce also has a right to health and safety information so that safety representatives or employees can carry out their functions. Where a trade union is recognised, safety representatives appointed under the Safety Representatives and Safety Committees Regulations 1977[aa] must be consulted by employers on matters affecting the group or groups of employees. If a trade union is not recognised, employees must be consulted by their employers under the Health and Safety (Consultation with Employees) Regulations 1996[bb]. The employer can choose to consult directly with employees as well as through directly elected representatives where they exist.

103. The transient nature of construction means that it has always been difficult to recruit and retain members and a site of 50 employees and above is the exception rather than the rule.

---

[y] http://www.hse.gov.uk/statistics/industry/construction/injuries.htm
[z] The Information and Consultation of Employees Regulations 2004, SI 2004/3426
[aa] The Safety Representatives and Safety Committees Regulations 1977, SI 1977/500
[bb] The Health and Safety (Consultation with Employees) Regulations 1996, SI 1996/1513

104. Genuine consultation involves taking account of employees' views and should take place before decisions are taken to give time for a dialogue to take place.

105. For effective worker representation and consultation on health and safety a number of conditions must be in place including demonstrable senior management commitment both to health and safety and a participative approach; together with effective autonomous worker representation at the workplace and external trade union support. Given the structure of the construction industry and the relative weakness of the trade unions, arrangements for consultation are weaker the further down the sub-contracting chain one goes. As Walters and Nichols stated in their case studies "*it was clear from these results that workers at the end of the chain of communication created by fragmented employment relationships on construction sites, fared least well in terms of provision of information and consultation and also rated managers least effective at managing health and safety*"[cc].

106. The views of employers and trade unions were in stark contrast when it came to the role of trade unions in promoting site safety. The trade union view is that where there is a trade union and safety representative it is a safe site. Research confirms this in other industries such as manufacturing although it may not be directly comparable as they are 'fixed' premises.

107. Trade union representation appears to be about 10%. One reason for this could be the anti-union stance of some employers but it is also a reflection of the general decline in trade union membership in the private sector. There is no doubt that the existence of a "blacklisting" agency and the ease with which the industry can dispense with the services of any construction worker let alone so-called "trouble-makers" does suggest at best a wariness of genuine, bottom-up worker participation and suggests the absence of the preconditions necessary to support effective representative participation for workers in large sections of the industry.

108. To extend this safety culture, some proposed roving safety representatives who could walk onto sites whether or not there were trade union members. Others thought there should be automatic trade union safety representative recognition if there were more than 50 employees on site. Not once did anyone refer to the European Directive[dd] on information and consultation.

[cc] Walters D and Nichols T. 'Worker representation and workplace health and safety', Palgrave Macmillan, Basingstoke, 2007.

[dd] EC Directive on Information and Consultation (Council Directive 2002/14/EC)

109. Most contractors I spoke to pointed to the importance of leadership and management of site safety which brought about culture change. Some companies are content to work alongside trade unions to promote this. Others pursue worker engagement policies which do not necessarily involve trade unions. There was a tendency by contractors to play down the role of trade unions on safety issues. Some companies expressed enthusiasm for worker engagement and behavioural change as a tool to promote a safety culture. It was unclear whether this was an attempt to fill a void in the absence of a trade union or a device to avoid union involvement. There may be a mixture of motives. Worker engagement is a legitimate management tool which is normally more highly regarded by managers than by workers when they are surveyed for their opinions. Successive Worker Employment Relations Surveys seem to support this[ee].

110. The suggestion of automatic rights for non-union safety representatives was mentioned as a possibility. Others were more sceptical saying it was difficult enough to find willing trade union representatives. One said "*I do not think they (non-union representatives) would come forward. Someone would be appointed without sufficient training. They might be self-employed. If you suggested someone was released to do health and safety training who was self-employed they would think you were on planet Zog*".

111. Many raised the issue of worker safety advisors which had been the subject of a pilot across various industries not just confined to construction. The Government provided funds to create these posts, selected mainly from trade unions, to act as advisors rather than representatives on health and safety in identified regions or sectors. The work was supervised by the HSE and an assessment was done on the project[ff]. The Government and the HSE saw worker safety advisors as a pump-priming project and long-term funding was not envisaged.

112. On the whole there was an enthusiastic response from the trade unions although some commented on the top heavy bureaucracy of the projects which made them more expensive than they needed to be. Some construction federations also mentioned the scheme with enthusiasm and they had continued to fund them for a period when the pump-priming money from Government ran out. Support for continuing or resurrecting the worker safety advisor project depends entirely on willingness to fund it. Many stakeholders supported the idea but expected someone else to pay for it. Various organisations are still working on the feasibility of continuing to develop worker safety advisors

---

ee http://www.berr.gov.uk/whatwedo/employment/research-evaluation/wers-2004/index.html

ff York Consulting. 'The Workers Safety Advisor (WSA) Pilot', HSE Research Report, RR144, 2003.

but it is generally acknowledged that one organisation alone cannot sustain it. Dame Carol Black, in her report[gg], recommended that trade union safety representatives, where present, should play an expanded role to promote the benefits of employer investment in health and well-being. She called for the development of "*a robust model for measuring and reporting*" on these benefits and indicated that employers should report on health and well-being in the boardroom and company accounts. I hope that any plans for taking the initiative in this area could include funding for some worker safety advisors in the construction industry.

113. I was given examples of sites where trade union involvement was regarded as acceptable and positive but these were in the minority, particularly away from the larger sites. I believe this is a missed opportunity by the construction industry.

114. I believe that engaging the workforce in a strong safety culture in the way that the oil or nuclear industries have, so that they are alert to risk and can speak out, would reduce fatalities and serious accidents, particularly when combined with leadership from the top and effective site management. Trained trade union safety representatives are a potentially powerful resource for improving health and safety and making a significant contribution to sustainable improvements in other sectors and are currently underutilised in construction. However, efforts should continue to be made in the medium and long term to spread this bottom-up approach throughout the industry, starting with the large construction companies who are already experimenting with such schemes. Trade unions have shown that they can do much to assist the promotion of good practice on large, high profile sites where they, the employers, HSE Inspectors and others have been engaged in planning occupational health and safety management across the whole of the supply chain. Research is needed better to understand how such practices can be used elsewhere in the industry.

115. On roving trade union safety representatives, I think it would be difficult, if not impossible, to confine such a development to construction alone (although this has happened in Norway) and I believe that no Government would be minded to initiate such a Regulation as a general employment requirement.

116. I do not believe that the positive role trade unions can play is sufficiently appreciated in the industry, not just at national policy or negotiating forums but in helping to build a health and safety culture amongst a

[gg] 'Dame Carol Black's Review of the health of Britain's working age population: Working for a healthier tomorrow', March 2008. http://www.workingforhealth.gov.uk/Carol-Blacks-Review/

workforce with diverse employment status. A more constructive acceptance of this by construction companies would be an important step forward.

Employment by Gangmasters

117. Although I made no specific recommendations about more legal rights for trade unions or safety advisors I do think that the floor of employment, safety and health practice in any industry should be underpinned by regulation. The Government has been very active in the area of employment agencies and employment businesses strengthening the role of the Employment Agency Standards Inspectorate (EASI) and increasing its numbers. This is a very commendable development but it has to be recognised that it remains a very small unit within BERR (now BIS[h]). Similarly the work of the Vulnerable Worker Enforcement Forum and the Fair Employment Enforcement Board is very welcome.

118. However, workers at the end of the chain on construction sites are likely to be the least protected on safety and other grounds. Traditionally the floor in the construction industry has always been regarded as the labour-only sub-contractors. It may still be the case that this is the most exploited area in the industry after the informal economy. One of the difficulties about this huge fragmented industry is obtaining reliable statistics and concrete evidence of abuses in its darker corners. Society appears to accept a more buccaneer approach to work in construction, which if applied to another industry or profession would lead to shock and anger.

119. The number of workers used by gangmasters for construction purposes, by the very nature of this twilight zone, is not known. Some estimate it is 3% of construction workers which in itself constitutes a very large number of people. It may be more. It may be less. Most of us have anecdotal evidence of streets or car parks where men wait around before dawn to be 'selected' to work for the day.

120. Some claim that gangmasters who have been forced out of areas covered by the Gangmasters Licensing Authority have moved into construction. I am not sure that matters either way. If they exist they should be licensed. If they are above board they have nothing to fear. Gangmasters provide a service which is valued by some and we need to differentiate between the good and the bad.

121. If this is the construction 'floor' it should be licensed. To date Government has not been minded to extend the remit of its

Gangmasters Licensing Regulations[hh] which were made to ensure that the events of the Morecambe Bay cockle pickers should never be repeated. However, I do believe that there is a case for doing this even though it requires primary legislation. I have been told by some stakeholders that it will be very difficult to differentiate workers taken on by gangmasters or by labour-only sub-contractors. I think this rather proves the point about the relatively unregulated end of the industry. If the resources of the Gangmasters Licensing Authority (GLA) are taken up with its existing responsibilities some thought should be given to reconstituting it and reviewing its resources to include construction in its remit. If this happens and unlicensed gangmasters then move into employment businesses or labour-only sub-contracting, it is vital that resources are provided to ensure that EASI can do its job, and that there is a continuation of a 'joined up' approach.

122. I accept that this represents an additional regulatory responsibility. Under the existing structure of the construction industry it has extraordinary flexibility when deploying its labour. It may be correct that very few contractors use gangmasters directly. It is less clear how many they may use through their supply chains.

123. We should be putting out the clearest possible signal that, as with the statutory National Minimum Wage, there is a floor below which a society should not tolerate exploitative practices. I firmly believe that extending the GLA's remit to construction or making an effective Regulation with the same objective would provide that signal.

Vulnerable Workers

124. There has been a tremendous amount of work done on the issue of migrant workers throughout Government, Whitehall, the trade unions and campaign groups. The comparison study in the Phase 1 report and the Phase 2 case studies reflect that concern. As has been pointed out, the number of migrant and foreign workers killed in 2007-2008 in construction increased from 8 in the previous year to 12 out of a total of 72 deaths.

125. Some argue that the number of migrant deaths proved they were more vulnerable because they were exploited, or were used to tolerating lower standards of health and safety or there was a language barrier which prevented effective communication on safety issues. However, we do not have the data to distinguish this hypothesised causal link from an equally plausible one that this increase results simply from a greater number of migrant workers being at work in construction in that year.

[hh] The Gangmasters (Licensing Authority) Regulations 2005, SI 2005/448

With the huge influx of migrant labour in recent years, many of them single young men with no skills, it is hardly surprising that they would flock to the construction industry which was experiencing a boom year in 2007-2008 and which on the whole did not make too many demands in terms of entry qualifications. This does not present the total picture, however. Many of the migrants were skilled and experienced and worked in regular groups. They were very attractive to the industry which liked their work ethic and they were paid the agreed industry rates. Some questioned the competence of the migrant skilled workers and said their qualifications were not comparable; some were concerned about the potential for industrial relations problems and/or racial tension; some felt that the large numbers of migrants working in construction allowed the industry off the hook when it came to investing in skills and training of the indigenous population.

126. While acknowledging the vulnerable position of some migrant workers, issues of vulnerability extend beyond migrants to include other groups. One could argue that anyone who does not have the competence or confidence to report unsafe conditions or practices, or to refuse unsafe work for fear of losing his job, should be seen as a vulnerable worker. That is not to say that individuals are not responsible for their own safety. Too many accidents and deaths are caused by complacency or cutting corners by individuals or the "*it won't happen to me*" attitude. There are many examples of fatalities where the worker had been warned about his actions or where they had actually written the method statement or had site supervision responsibilities. Too often there is a general culture which is not conducive to team work or looking out for each other. Constant site awareness or ignorance of the risks of unfamiliar tasks were often cited in the case studies as factors. However as one stakeholder commented "*A temporary lapse should not kill you. We should not accept that construction has to be a dangerous industry*".

127. Strong leadership can and does change cultures and lead to the introduction of newer and safer technologies and work methods and there have been some examples of a remarkable turn around in health and safety attitudes which have led to a drop in accidents and fatalities e.g. by implementing the Incident and Injury Free Programme (IIF). The Department for Work and Pensions' Job Centre Plus building programme delivered before time and under budget with no fatalities and is often cited as a model. However, where strong leadership does not exist and therefore site conditions and supervision are poor, all construction workers are vulnerable.

128. Two groups have been mentioned by stakeholders as particularly vulnerable – the very young and the 55-60 plus experienced worker. The very young worker might be an apprentice on an unsuitable site or a

casual labourer with no skills and experience, ignorant of their rights and obligations, and an unwillingness to challenge authority or question an instruction.

129. The 55-60 plus worker may have been in the industry a long time. Stakeholders commented that this group was sometimes poor at getting a health check either for eyes or ears or, as one commented, "*their backs might be on their last legs*". Some might underestimate the changes or complexity of equipment e.g. dumper trucks. One commented "*men over 50 doing the job for twenty years take risks. A frightening number fit this mould. People who do things really daft. It does not matter how much paperwork or directors' campaigns you do*".

130. There may be some useful work to be done in targeting specific groups of workers for safety campaigns to raise awareness or act as timely reminders of dangers at work. There might also be some useful research to be done on exploring further whether any other groupings or sub-groupings of migrants, the young and the 55-60 pluses are particularly vulnerable. In Paragraphs 117 to 123, I have also made recommendations about those who are taken on by gangmasters.

OCCUPATIONAL HEALTH

131. Although the Inquiry was asked to concentrate on fatal accidents in the construction industry, it needs to be recognised that far more workers die from the chronic effects of ill health caused by or made worse at work. 20 tradesmen (electricians, plumbers etc) die every week from asbestos related disease. Research[ii] indicates that a further 12 construction workers die every week from silica related lung cancer. While these deaths relate to past exposure, this is a continuing problem where workers can be at risk now and their health is affected in 15-40 years' time.

132. It is difficult to compare directly fatal accident data with deaths resulting from occupational ill health. It is often difficult to say exactly when and where potentially fatal exposures to hazardous substances have occurred and this is reflected in the lack of specific data collected:

Ill Health	Accidents
• **2056** people died of mesothelioma (2006), and thousands more from other occupational cancers and lung diseases[jj]. *There are estimated to be around 4000 deaths/year from exposure to asbestos, many of these deaths are associated with workers from the construction industry.* >**500** people died from silica related lung cancer in 2004[kk]	• **72** workers were killed at work in construction in 2007-8[jj]
• **1.7** million days were lost in 2007-8 due to work related ill health, which equates to 0.77 days per worker[jj]	• **1** million days were lost in 2007-8 due to workplace injury which equates to 0.45 days per worker[jj]

133. The construction industry is one of the worst performing for occupational health. Data from the respiratory disease THOR GP surveillance schemes[ll] for 2006-2007 indicated a higher incidence rate in construction than across all industries (although the rate of self-reported work-related ill-health in construction is of the same order compared with all industries). Construction also has the same or above all industry average incidence rates in a number of specific areas:

---

ii Imperial College London and the Health and Safety Laboratory. 'The burden of occupational cancer in Great Britain', HSE Research report RR595, 2007. (The table reproduced in Paragraph 132 is Figure 14, page 91 of the main report)

jj http://www.hse.gov.uk/statistics/industry/construction/index.htm

kk COHME website: http://www.hse.gov.uk/construction/healthrisks/

ll THOR GP, see: http://www.medicine.manchester.ac.uk/oeh/research/workrelatedillhealth/sicknessabsence/thorgp/

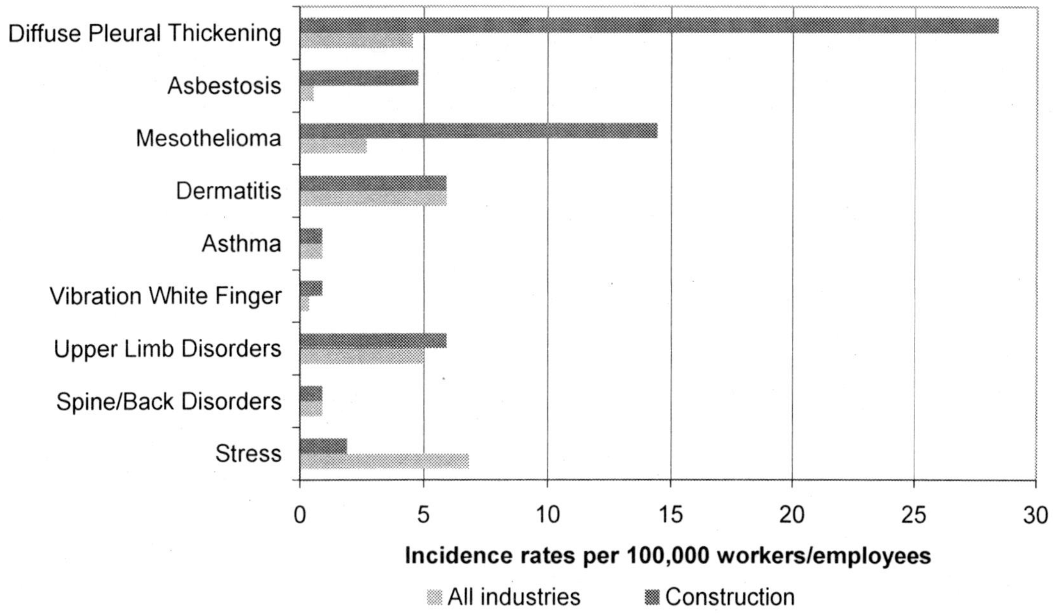

134. Occupational cancer deaths amongst construction workers and tradesmen reflect the risks inherent within particular construction work processes, environments and materials. However, they also reflect a failure by the industry to prevent or control exposures and so adequately manage this issue. For too long, health has had minimal attention when compared with safety. Thus, while significant progress has been made by the industry on safety issues during the last decade, it has failed to achieve the same for ill health.

135. Within this general picture it is also important to note the disparity between large contractors and the significant number of small and medium sized enterprises (SMEs) making up the rest of the industry. While some larger companies have embraced the importance of tackling ill health issues it is often a matter of last resort for SMEs who are more focussed on the necessity to 'make do' and get the job done. For this group sometimes even the provision of adequate temporary welfare facilities proves a step too far.

136. To secure improvement, the industry needs to manage both health and safety issues as an integral part of its day-to-day business management – to manage the risk not the symptom – to paraphrase an HSE message.

137. HSE has produced the Construction Occupational Management Health Essentials (COHME) website, which provides an advisory resource of safer working practice for the industry. The Constructing Better Health (CBH)[mm] programme is actively working toward raising the standards of occupational health provision in the industry.

[mm] Constructing Better Health website: http://www.constructingbetterhealth.com/

138. There is clearly scope for more to be done to reduce the fatalities from chronic ill health, and whilst this is outside the scope of this Inquiry, I think it is essential that the construction industry increases its efforts to tackle this issue. The risks are known and it is unacceptable that the number of deaths continues to be so high. Some of the causes of the fatalities from chronic ill-health may have been from contact in the last 10 or 15 years so we have little to be complacent about. The scale of the problem is significant and, it would appear, is the result of the many of the same structural conditions of the industry that lead to the poor performance on injuries. Tackling these problems in the long term would seem likely to reduce the incidence of both fatal accidents and serious injuries and to improve occupational health.

COMPETENCE / TRAINING

General

139. The history of skills training in UK construction could cover several volumes. All the major reports refer to the industry's poor image and reluctance of parents to encourage entry into the industry as one of the major draw-backs. There is also a conflict between building a world-class skills base and providing job opportunities for people who are unskilled and find it difficult to obtain employment. Some in the industry think it is possible to achieve both. Time and again stakeholders would indicate that the main problem with the industry is that there are no entry requirements. That in itself should not matter. The Open University, celebrating its 40th anniversary this year, has no barrier to entry either. The difference is what you do with the entrants to enable them to develop and fulfil their potential once they are admitted. As the majority of the workforce (approx 55%) have skills below NVQ level 2 or equivalent and approx 11% hold "*low or no qualifications*" (the Business and Enterprise Committee report[x]), it is clear that there is still a long way to go. The financial structure of the training industry means that it is easier to poach from other companies than train themselves. That is a problem which is not exclusive to construction.

140. Looking at the fatality case studies what emerges is that an almost equal number of qualified but inexperienced and experienced but unqualified workers were involved. So both elements are vitally important.

141. Some of the barriers to progress mentioned by stakeholders are as follows:

    a) The failure of the majority of construction companies to take on apprentices

    b) The limited number of apprenticeships but particularly the high drop out rate

    c) The inappropriateness of some of the courses

    d) The failure by some public sector clients to insist on CSCS cards as a minimum requirement when considering tenders

    e) The ready supply of migrant labour

    f) The cost of training if an individual was self-financing.

142. ConstructionSkills (formerly known as CITB) provides a vital role in providing grants for training from the long established levy system. I am not sure whether it performs a major redistributive role as the larger companies have the resources to maximise their opportunities and the smallest companies on the whole would not be able to afford to release

people to attend courses. According to statistics provided by ConstructionSkills' Training arm, 1425 small companies claimed a grant in 2008 (compared with 1172 in 2007) out of a total number of companies of 3150 (2558 in 2007). On the face of it this is an impressive example of redistribution. However, their definition of a small company aligns with the European definition which is 50 employees or less. In the construction industry in the UK a figure near 50 employees would be regarded as a fairly large company. Figures for micro companies (less than 10 employees) claiming grants in 2008 were 678 (475 in 2007) which may give a clearer impression of how small companies relate to ConstructionSkills. It has recently been announced that since the Inquiry started health and safety grants to the construction industry have been reduced by ConstructionSkills which is a matter for some regret.

143. Nevertheless it should be emphasised that the work of ConstructionSkills, and in particular its recent development of improved site management training in conjunction with the UKCG, plays an important role in improving the competence and health and safety awareness in the construction industry.

144. The development of the CSCS cards by the industry and the continuing commitment of the UKCG and the SFfC to its success has been an important development in the industry. CSCS has been administered by ConstructionSkills on behalf of CSCS Ltd (a partnership Board with various industry representatives) and changes are taking place in the running of the system, not least the proposal to produce time limited smart cards and merging CSCS Ltd with Constructing Better Health – an organisation which specialises in occupational health.

145. There are inevitable tensions between an organisation which promotes competence (ConstructionSkills) and one which records skills (CSCS Ltd). There were many critics of the CSCS cards and their views could be summarised as follows:

a) Need to reduce fraud.

b) Health and safety test is too easy.

c) Should be a ticket to carry out the work not just an entry to work on the site.

d) The level of skills on a CSCS card is often totally unrelated to the level of work carried out on site.

e) CSCS Ltd were trying to diversify before they had consolidated their existing work.

f) CSCS was not suitable for certain specialist areas.

g) CSCS Ltd does not engage with industry, even its supporters.

146. Whatever the justification of these criticisms, it is clear to me that the CSCS cards remain an important part of improving health and safety awareness on site. Although the system has been established for more than 10 years those involved with CSCS appreciate that it needs to become more strategic and needs to be relaunched with renewed vigour. The introduction of a smart card may help to reduce fraud and it may be necessary to review the standard of the touch screen health and safety test. Some major contractors accept that it is only a first step to improving the industry. I believe the consolidation and improvement of the CSCS system should be strongly supported by all in the industry and that Government should renew its insistence that all publicly procured contracts should make the possession of CSCS cards a condition of appointment.

147. Turning to the issue of apprentices, there is clearly a large number of applicants for every vacant place (one figure showed six applicants for each place) so there is no lack of interest despite the difficulties of finding, and then staying on, suitable placements. However the drop-out rate is also extremely high. Different reasons were given to me such as young people did not have the commitment or skills to complete the course or that young people received a poor deal and were taught on inappropriate courses or were only trained for 16 hours a week instead of 33 hours which applied years ago. One stakeholder commented that apprentices training had gone back to the 1950s. The picture is slightly better in Scotland with eight apprentices for every 100 workers compared with 0.9 for every 100 workers in London. The extent of self-employment must have an impact on training. I was told that 94% of painters and decorators in London were self-employed which explained why colleges dropped their painting and decorating courses.

148. The issue of good apprentice training for appropriate numbers of young people is important to the Inquiry for two reasons: A trained workforce will reduce fatal accidents and young people are more vulnerable on site if not subject to suitable supervision. The point was made by some stakeholders that the narrowness of training sometimes adds to the risks on site where a multidisciplinary approach was often expected. This led to workers performing tasks with which they were not familiar and this was often an underlying cause of an accident. The example was given of builders who installed solar panels (which supported the sustainability agenda), but who were not skilled roofers and often put themselves in danger. Another point was made that the 16-19 age range for apprenticeships was not appropriate for certain trades e.g. sheeting and cladding where a more mature approach was required. A more flexible approach to apprentice training and grants and a more targeted

approach to certain specialisms would improve retention and qualification rates. I am aware that there is a lot of work being done in this area by the Department for Innovation, Universities and Skills (now part of BIS) and ConstructionSkills so I will limit my comments to emphasising the importance of getting apprenticeships right and suggesting that further work is done to ascertain why the drop-out rate is so high from courses when the level of applications is also high.

Graduate / Post Graduate

149. It is not my intention to cover in detail the area of professional and postgraduate training in higher education or enter the debate about specialisms versus multi-disciplinary education or the need for more research and development in construction.

150. However underlying causes of fatal accidents are often to be found in design, leadership and planning and it is important that we ensure that sufficient consideration is given to these in our university courses. It is recognised that some site accidents could have been prevented by better appreciation and elimination or mitigation of hazards at the design stage. This should apply to the built environment and to the design of construction machinery and equipment.

151. Many stakeholders raised the issue of insufficient awareness in the university curriculum of health and safety or of the revised CDM 2007. If these subjects were covered on the curriculum were they only optional add-ons or an integrated part of the course? Was there sufficient linkage between what is taught about building and project design and methods on the one hand, and causal factors of accidents and fatalities in building and maintenance on the other? Were the assessors in professional courses sufficiently aware of the importance of CDM Regulations, and risk assessment? If risk management and health and safety were on the curriculum, how were they monitored to ensure sufficient coverage and proportionality?

152. Latham talked about "*an acceptance that a greater interdisciplinary approach is necessary, without losing the expertise of individual professions*". Egan called for the right balance of technical and leadership skills at top management level, the development of training programmes for project managers "*in integrating projects and leading performance improvement, from conception to final delivery*". Finally Egan called upon designers to match their high standards of professional competence "*by a more practical understanding of the needs of clients and of the industry more generally. They need to*

develop greater understanding of how they can contribute value in the project process and the supply chain".

153. It is clear that the major contractors hire a considerable number of management trainees each year most of them with construction related degrees. In fact they appear to far outweigh the numbers of apprentices they take on.

154. I received anecdotal evidence that one university had offered a postgraduate course in construction management without making any reference to the CDM Regulations and one Professor of Architecture (admittedly about ten years ago when the original CDM Regulations had been in existence for only six years) who had never heard of the CDM Regulations. One stakeholder made the point that some architects and designers still see themselves as clients' representatives or even the client but not part of the construction team. One CDM Co-ordinator said "*some (designers and contractors) ring up a day or two before work is due to start and expect CDM Co-ordinator's input. Half the phone calls are like that. Designers haven't a clue. They are not engaged with the safety element*".

155. I am not aware of any current review between HEFCE (or the equivalent bodies in Scotland and Wales), the industry and professional bodies on the adequacy and relevance of the university or college curriculum on undergraduate and postgraduate construction related courses.

156. Some work was carried out for the HSE in 2001[nn] and followed up in 2004[oo] on this area. The first report concluded that "*although isolated examples of excellence existed, both accreditation bodies and universities / schools of architecture had much to do in terms of setting adequate standards, and integrating health and safety into the curriculum*".

157. I understand that a high proportion of students on undergraduate and postgraduate courses are from overseas and are an important source of revenue to universities. Many have every intention of returning to their own country to work and would not require knowledge of a country specific or European set of Regulations such as CDM 2007. I do not think that this should be allowed to be an opt-out clause for universities and colleges when reviewing whether the four points set out in Paragraph 160 below are adequately covered. Risk management and

---

[nn] Symonds Group Ltd, Liverpool John Moores University and Charlton Smith Partnership. 'Identification and management of risk in undergraduate construction courses', HSE Contract Research Report CRR392/2001, 2001.

[oo] Symonds Group Ltd, Liverpool John Moores University and Charlton Smith Partnership. 'Identification and management of risk in undergraduate construction courses – Supplementary report', HSE Research Report RR275, 2004.

the importance of project team-working, pre-planning and health and safety are important elements wherever the qualified student decides to work.

158. The research report[nn] recommended that "*those accreditation bodies that currently have minimal reference to health and safety review their requirements, and that all accreditation bodies actively work towards a pan-industry approach*". It also recommended that health and safety should be taught as an integral part of construction risk management and that an industry umbrella body such as the Construction Industry Council leads in developing a pan-industry, standard curriculum template for health and safety risk management delivery in undergraduate construction courses.

159. It would be interesting to have a tripartite review of these issues five years after the report up-date and ten years after the 1999 'Teaming Up for Education' conference[pp].

160. I would urge that, if it is not already being carried out, some work is done to ensure the following in universities and colleges:

    a) Health and safety is appropriately covered in the curriculum including design and maintenance implications and site awareness not just technical and legal implications.

    b) CDM 2007 are appropriately covered so that risk assessment and risk management are fully comprehended for people on site as well as those in the built environment.

    c) Assessors are fully aware of developments in health and safety awareness and CDM Regulations.

    d) Moderators should ensure that the above work is carried out in practice.

[pp] The 'Teaming up for Education' conference was held in 1999 jointly between HSE and the Construction Industry Council (the representative forum for the professional bodies, research organisations and specialist business associations in the construction industry), and brought together representatives from universities, professions and the industry to discuss ways of integrating health and safety into construction-related courses.

BEREAVED FAMILIES AND THE LEGAL SYSTEM

161. The stark reality of the impact of a construction fatality was nowhere more obvious than when I met the families and representatives of the deceased on different occasions throughout the Inquiry. I experienced conflicting emotions which I am sure are not unusual in that I was forcing the families to re-live their grief but at the same time I needed to know their view of how the industry, the HSE, the courts and society as a whole dealt with them. I am extremely grateful to them for being prepared to see me. Some were sceptical that anything would happen because they did not see any sign of improvement in the 'system'. Their experience was very similar, their dignity impressive and their proposals were centred on future prevention rather than vengeance or compensation.

162. Each one in different ways, reflected the comments of all:

- *"I don't want this to happen to anyone else. I don't want families to go through what we are going through."*
- *"It is not about vengeance. It is about getting it stopped. If the Government doesn't do something about it, it will go on and on."*
- *"I brought up my sons to be law-abiding. You go out to work for the things you want. They were killed by someone more interested in money.... People think they are safe at work. Sometimes I would have to get him up at 4.30am."*
- *"There should be more unannounced inspections. They always know when an inspector is coming. When you have got a visitor, you clean your house."*
- *"The CPS does not understand health and safety law, nor do the coroners. Magistrates and judges do not understand it either. You have this horrendous thing to deal with – the death of your son – and then you have to fight every step of the way."*
- *"There is complacency that someone has been killed. You have a battle. The trade unions are carrying the burden. Most people cannot afford it. You get no help, no counselling. You are on your own."*
- *"No-one is responsible. The paperwork is there, the method statement is there but not acted on. The focus has gone from the practical. Nothing will change until the responsibility chain is tackled. Everyone was responsible for every bit of the site except the bit my husband was on. I'm sure the responsibility would have moved if it had been a different bit of the site. 350 men have died since my husband died. In any other profession there would be an outcry."*

163. I have dealt with the issues of responsibility and the social unacceptability of construction deaths elsewhere in the Report.

164. On the issue of inspection by the HSE, every bereaved family member remains absolutely convinced that an inspection visit is known and expected by site workers. Equally HSE Inspectors are adamant that this is not the case and emphasise that they sometimes do not know themselves which sites they will visit until the day of the visit. This could be because many visits of inspectors are after accidents, when it is of course more predictable. No amount of investigation on my part has identified reasons for this polarisation and this might be the subject of a useful study.

165. On the issue of delays in the legal process, there was no common ground. Some blamed the coroner's system which is underfunded and lacks consistency. The new Bill[qq] may help to provide some consistency but unless the system is properly resourced the delays will not improve substantially. Although coroners will eventually be full-time, part-time coroners (who are the most under-resourced) will continue to work and will only be phased out through natural wastage. So any changes in the system are likely to be very slow.

166. Some held the HSE responsible for delays in both investigation and in reviewing a case after an inquest or after the police have referred it back. The HSE indicates that it bears the brunt of the blame because it comes at the end of a long and convoluted process. Some blame the police for concentrating on the 'smoking gun' aspects of a fatality rather than the underlying causes and for not always appreciating the seriousness of health and safety legislation. Some feel that the Crown Prosecution Service, the Magistrates and the courts do not fully appreciate the importance and urgency of dealing with a construction fatality. In Scotland a specialist division is being established in the courts for dealing with these fatalities and it will be interesting to see how this develops.

167. At one court case I attended which had taken five years to conclude, the judge commented "*somebody needs to address these long delays. We are on our fifth HSE Inspector. In essence it is not a complex case. I wish we could find a better procedure*".

168. I have had insufficient time to have detailed discussion with all the authorities involved in the legal process to make an informed comment. I am unclear whether these delays are particular to construction or if they are typical of the system or due to lack of resource. What is clear is that

---

[qq] Coroners and Justice Bill 2008-09
http://services.parliament.uk/bills/2008-09/coronersandjustice.html

the delays are unacceptable and that additional resources are unlikely to be made available in the near future. The 'system' appears to have built-in delays whereby a time lapse at any or all of the stages leads to years of stalemate. If the construction company as defendant decides to fight a case vigorously or there is delaying action by solicitors, this also adds to the delays and it is difficult to see how families can possibly start to recover from their loss let alone endure the lengthy legal proceedings. I understand that protocols exist between HSE, coroners and police to try to break the log-jams in the procedures. I believe there needs to be a thorough piece of work carried out to identify the in-built delays and to recommend improvements.

169. On the issue of representation and support for families, I can only recommend that a construction worker should be a member of a trade union, or if they are self-employed, have sufficient insurance to enable their families to obtain legal advice should it prove necessary. Guidance to bereaved families does exist but it does not seem to be readily available or comprehensive. Most construction companies would not even hold the names of close family contacts in the event of accidents.

170. Not all families want to prosecute as they sometimes accept that the deceased had contributed in large measure to his own death, or that the employer or partner responsible for the work on site was a life long friend or they wished to reflect what they thought the deceased would want. However, I feel, without conclusive evidence, that some legitimate claims are not pursued either through lack of knowledge or financial support.

171. While the delays in the legal system and the way in which bereaved families are treated are not, in themselves, underlying causes of fatal accidents, they contribute to the way in which society regards these tragedies. Where a case is pursued, the system ought to provide a timely and proportionate conclusion.

PRE-QUALIFICATION

172. As I have indicated elsewhere the structure and practices in the construction industry add to its complexity and, in turn, divert attention away from the practical application of health and safety. This is evident in the system of pre-qualification questionnaires and other pre-qualification practices.

173. The principle of pre-qualification is a good one as it is intended to encourage and reveal readily available information about a company's standards, competence and suitability. Latham[b] described it as follows: "*'Prequalification' is an effective system which allows clients to seek tenders from contractors (and also consultants) of equivalent size, capability and experience. Strictly speaking, this is a two stage procedure. 'Qualification' means a contractor getting on to an approved list at all. 'Prequalification' means drawing up a list of firms which are suitable for a particular project. The first stage is the necessary gateway to the second*". He then referred to the number of Government Department and local authority lists of contractors in existence at the time and went on to say: "*such duplication of effort is a wasteful burden for the construction industry. It also adds to clients' costs, not only because of increased prices from firms but also because of the resource implications of maintaining separate lists. Attempts to rationalise the procedures in the past have not met with success*". He recommended: "*a single qualification document for contractors seeking to do work for any public sector body. This will need to take account of any special interests particular public bodies may have. The system of Public sector Project Managers devising their own pre-qualification forms should cease*".

174. Although action was taken to devise a public sector list which still exists as Constructionline (run by Capita), it did not prevent the proliferation of pre-qualification systems. In my meetings with stakeholders it became clear that there were at least two separate groupings trying to find means of agreeing common bench-marks with very little evidence of success. The obstacles to progress appear to be lack of political will, loss of potential revenue by the pre-qualification schemes and a tendency in the industry to say "*not invented here*". One stakeholder commented that there was probably £5m of professional time spent by contractors on pre-qualification which is almost certainly a conservative estimate. Some, it is claimed, are no more than paper exercises to enable a client or their representative to tick the box that a contractor has been tested. Some schemes are extremely demanding and can only be met by companies offering the highest standards.

175. CDM 2007 put competencies centre stage[rr]. This means that the rationalisation of core competencies is more urgent than ever.

176. One stakeholder felt there was a question mark over the competence of all the competency schemes. Business was the main driver, not the assessment. "*The client should have a health and safety focus but it is not happening*". What is clear is that health and safety is sometimes weighted as one of a list of requirements instead of being an absolute requirement. Although Latham recommended one list for the public sector, I do not believe there is sufficient motive or leverage in the industry to make it effective.

177. Health and safety, along with financial viability and capacity, should be absolutes in any core competency considerations. They should be yes/no and not weighed against other criteria. In addition there are many criteria for judging the safety and health competences and the different schemes do not agree on them. The issues is not so much about the number of schemes as finding standard bench-marks within those schemes so that sub-contractors are not having to conform to a host of separate pre-qualification schemes if they want to participate in tendering for a local school, hospital, housing association, or local authority project.

178. I have no doubt that if the money and time devoted by contractors to satisfying the myriad pre-qualification schemes were diverted to good health and safety practices on the ground, the construction industry would be a safer place. Major clients in the public sector must bear some responsibility for this bureaucratic nightmare.

179. I accept that the client is king but too often the pre-qualification schemes are used as part of a 'whittling down' process to make tender processes more manageable, or as a barrier to entry. The length of the tender lists has been an issue between client and contractors / professional groups for decades. The industry maintain, quite correctly in my view, that they would be more efficient if tender lists were limited. The client continues to prefer to maximise the opportunities for competition sometimes at the expense of health and safety standards. One stakeholder representing clients thought that long tender lists were linked to the knowledge and capability of the client. Large parts of the client community believed in lowest cost tenders and did not see the bigger picture. He wondered if there was a causal link between fatal accidents and the 'one-off' client who did not know where to go for good advice.

[rr] Health and Safety Commission. 'Managing health and safety in construction. Construction (Design and Management) Regulations 2007. Approved Code of Practice', L144, 2007. Paragraphs 202-212 – How to assess the competence of organisations and Appendix 4 (Competence) – Core criteria for demonstration of competence pages 109-113.

180. I recommend that the groups considering common core competencies for pre-qualification should try to co-ordinate their efforts and reach agreement on acceptable transferability in key areas. Health and safety should be an absolute in all common standards requirements. Agreement should also be reached on the core competences for health and safety. The Government, as a major client in public procurement, should take the lead in achieving one interlinked set of agreed standards on safety assessment in pre-qualification to cover the variety of needs.

181. On a separate point, some stakeholders suggested that a licensing scheme should be established for individual companies to show that they had the quality and capacity to undertake certain work. On the surface this sounds like an attractive proposition. However, I believe the conditions for a successful system do not exist. It would require stability in commissioning work and in the workforce so that the licence remained valid for a proportionate period of time. There would also have to be different types of licence depending on the specialism. What would constitute a construction licence? Would it apply to general builders and those driving on site? The construction industry is not short on bureaucracy and I believe this would add another layer without necessarily improving standards.

ROLE OF GOVERNMENT

182. The role of Government in public procurement and machinery of Government issues has been central for many years. Latham stated *"Government Departments – and the wider public sector – should deliberately set out to use their spending power not only to obtain value for money for a particular project but also to assist the productivity and competitiveness of the construction industry, and thereby obtain better value for money generally in the longer term"*. He went on to say *"Encouraging continuous improvements in productivity should be the driving force behind Government action as client, and the formulation of best practice should complement and contribute to this.....One Government Department should take the lead to ensure that best practice and the drive for improvement are implemented throughout the public sector as a whole"*. In 1994 this was the Department of the Environment.

183. Egan also emphasised the importance of the public sector as client and the role of Government: *"The task force recommends that the Government commits itself to leading public sector bodies toward becoming best practice clients. We believe that this process must begin with substantial improvements in the way that the public sector procures construction. In our view this can be achieved while still meeting the need for public accountability"*.

184. The Business and Enterprise Committee report 'Construction Matters' July 2008 referred to the number of Government Departments which have a strong policy interest in construction as client, regulator and provider of funding (paragraph 25[x]). This totals eight Government Departments not counting Departments such as Transport, Health and Defence who are all major clients. BERR at the time of the Inquiry and now BIS[h] takes overall lead for Government relations with the industry.

185. Given the complexity and significance of the industry it is clear that different Government Departments will want to retain their particular interest and having one Department co-ordinating all three roles (client, regulator and provider) is unlikely to be achieved and may not be desirable. It is important, however, to have more political focus on the industry and a resource to provide an overview of the various departmental activities. This is not intended to be a criticism of the Minister at the time of the Inquiry who worked hard to maintain the construction brief but I believe that part of ensuring that construction fatalities become socially unacceptable is to raise the profile of construction in Government. This would go a long way to convincing the industry and the population at large that construction is important. It is more than an industry. It is about the built environment and how we live

in it. Although there has been a great deal of activity on construction related issues by the Government – as the Inquiry was progressing there were debates in Parliament about the Coroner's Bill, Gangmasters, Blacklisting, Skills and Apprentices, Payment Schedules in the Industry, Health and Safety, Corporate Manslaughter to name but a few – there needs to be more focus and a higher status in Government. Despite the reservations of the Business and Enterprise Committee, I do believe there should be at least a full-time Construction Minister, with a co-ordinating brief to lead on construction. It should not involve "*a significant reorganisation of the machinery of Government*" – a matter which concerned the Committee. Granted such Ministers will move around Government Departments and not necessarily stay long in the Construction Brief, but they will acquire a body of knowledge which will not disappear and would help to create more understanding of the industry.

186. The debate about whether there should be a Chief Construction Officer[x] would then be part of any ministerial development. While supporting the general principle of having a Chief Construction Officer, there is no proof that this post would provide the long term continuity a Minister would lack. If the Chief Construction Officer had the kind of attributes required by the Business and Enterprise Committee – an in-depth knowledge of how the industry functioned, wide private sector experience, understanding of the workings of the public sector, commanding respect and trust of the industry and having sufficient influence in the Government – the likelihood is that this would be a pre-retirement post without necessarily providing long term continuity. A consultative document[ss] has been issued about whether the post should be created and if so, what functions it should have.

187. It is important that there is some clarity and focus to the position if it is to be created. I have doubts that one position, no matter how experienced and respected that occupant, could fulfil the role of champion, gatekeeper, co-ordinator and Whitehall monitor, all at the same time. The important point is that there should be a resource to make the construction role effective and to support the Minister for Construction.

[ss] BERR. 'Chief Construction Officer – Discussion document', February 2009. http://www.berr.gov.uk/files/file50091.pdf

PUBLIC PROCUREMENT

188. The important role of Government and the public sector generally in setting standards for procuring contracts and helping to raise standards in the construction industry has been a subject of discussion for many years.

189. The public sector as a whole procures 30% - 40% of total construction output in the UK each year. Government supports the view that public sector bodies should act as exemplars in procurement of contacts. The Office of Government Commerce (OGC) published guidance[m] entitled 'Achieving Excellence in Construction Procurement' on all aspects of the procurement process. Based on the OGC guidance Ministers approved mandatory key minimum procurement standards[n] – 'Common Minimum Standards; for the procurement of built environments in the public sector' (CMS). These standards include the need to develop long-term collaborative relationships between Government as client and its suppliers, use integrated supply teams, reduce financial and decision-making approval chains, adopt performance measurement indicators and use tools such as value and risk management and whole life costing.

190. The Common Minimum Standards also include standards on health and safety – for clients to:

    a) Have systems in place to collect performance data.

    b) Assess the performance and processes of supply teams as part of the pre-qualification process.

    c) Require their constructors to be registered with a scheme such as the Considerate Constructors Scheme and comply with the scheme's Code of Considerate Practice.

    d) Require all members of their supply teams to be registered with the Construction Skills Certification Scheme or prove competence in some other appropriate way.

191. As I indicated in Paragraph 180, I believe the Government should build on these minimum standards to arrive at a broadly accepted pre-qualification standard or set of standards for health and safety.

192. It is mandatory for central Government Departments (including their executive agencies and non-departmental public bodies) to comply with the OGC's Common Minimum Standards. However it is clear that some Departments are not providing sufficient evidence to prove that they have adhered to the Common Minimum Standards or that they have insisted upon their adherence lower down the chain. Despite the good work done by the Public Sector Construction Clients Forum chaired by

Dame Mavis McDonald the accountability is patchy. There seem to be some contradictions in the system which is probably unavoidable in such a complex area. The Common Minimum Standards are mandatory and the OGC guidance is advisory. Public money is being spent which demands accountability but the autonomy of various institutions is respected to the extent that the mandatory rules are not enforced.

193. The Business and Enterprise Committee report 'Construction Matters'[x] made a number of recommendations for further action including up-dating and promoting greater awareness of the Common Minimum Standards and making it mandatory for local authorities. I would certainly support this particular recommendation. Similarly the Strategic Forum for Construction's second report[tt] to the DWP Secretary of State emphasised that health and safety outcomes can be positively influenced by Government and local authorities taking an exemplary position on health and safety procurement. Lastly, the Better Regulation Executive's report 'Improving Outcomes from Health and Safety'[uu] has highlighted the need to ensure that SME contractors have an equal opportunity to bid for publicly procured projects. The report recommends that public sector procurers should have a presumption that SME bidders who are members of any health and safety pre-qualification scheme should be deemed to meet the procurer's requirements. I have commented separately on the pre-qualification issue in Paragraphs 180 and 191.

194. I was given many examples of projects which followed the OGC Common Minimum Standards and where the project was completed on time and within the budget as well as safely. I was also given many examples of public authorities selecting on price alone and tolerating lower standards on site. One argued against insisting on CSCS cards for all workers on the grounds that it would discriminate against builders in the local community and they had an obligation to consider benefit to the community when tendering for a project.

195. At national level the Public Sector Construction Clients Forum is very committed to health and safety on projects but suffers from inconsistent data from Departments and difficulty in tracking the very long tail to where projects are actually built. About 80% of the spending takes place at local level and information on health and safety is held at local level and rarely surfaces at national level. Where framework partnerships have been established there appears to be a better picture. These came into being between Government, client and major contractors to promote

---

[tt] Strategic Forum for Construction Health and Safety Task Group. 'Health and Safety Performance in the Construction Industry - Second Industry Report to HMG. Response to the Construction Forum: Framework for Action', August 2008.

[uu] Better Regulation Executive. 'Improving outcomes from health and safety', August 2008. http://www.berr.gov.uk/files/file47324.pdf

public sector building projects following Egan and the OGC guidance. At the time public sector building projects had a reputation for not being completed on time or on budget. One of the objectives of the framework partnerships is to bring best practice throughout the supply chain. The Government Department officials might help with process, procedures and contract management. Health and safety would not be their primary responsibility. However having looked at the Department of Health's ProCure21 projects it seems clear that their safety record is better than average.

196. It appears that the further down the chain the public procurer is, the less likely that OGC standards will be adhered to. This cannot be acceptable. The Audit Commission has identified that £2.5 billion a year could be saved on value for money in public procurement and health and safety procedures do not conflict with this. This is not a new problem. By insisting on strict adherence to the Common Minimum Standards at all levels of public spending and on a workforce with basic health and safety training, the taxpayer would gain and there would be fewer permanently injured workers relying on state support and fewer bereaved families.

197. Public procurement is important because of its size and its potential for insisting on driving up standards. The Common Minimum Standards are too often ignored at local level and there are few consequences when that happens. Those who are monitoring the effectiveness of the OGC Common Minimum Standard – the OGC itself and the Public Sector Construction Clients' Forum – should be supported in their endeavours. This can only be done by greater accountability of how tax payers' money is spent and effective means of identifying those who do not follow the mandatory standards. Many Ministers have tried to draw attention to the importance of standards in public procurement but without incentives or sanctions, it seems unlikely that further progress will be made.

198. I recommend that an effective monitoring system should be put in place to ensure that the mandatory OGC Common Minimum Standards are adhered to throughout the publicly funded chain and that effective sanctions are put in place for those who do not follow those standards or who do not provide sufficient accountability that they have followed those standards.

WORKPLACE EQUIPMENT / MACHINERY

199. As the Loughborough University[f] study has underlined, when international stakeholders were questioned about causes of accidents, the majority responded in terms of categories of 'accident mechanisms' (e.g. falls from heights) rather than about underlying causes (e.g. lack of knowledge). The Loughborough team thought that this *"possibly demonstrates the mindset of the international construction community towards a particular focus"*.

200. Although the Inquiry has concentrated on underlying causes of fatal accidents it should be recognised that historically fatalities have also been reduced through step changes in equipment and technology.

201. It is accepted that much effort is being made to improve site management and promote behavioural change at the top of the industry. However, it is important that focus remains on developing and refining technological support in construction.

202. Examples include the introduction and use of fall arrest nets, mobile elevated work platforms (MEWPs) and remotely controlled equipment for demolition. There was considerable resistance to their use for a couple of years mainly because of the cost of buying the equipment but their use is now commonplace and the variety and versatility of equipment has increased considerably.

203. Despite considerable advances in technology in the construction industry over the last ten years, it is clear that workplace equipment and machinery are still involved in a number of fatal accidents. Examples of two recent 'hot-spots' are tower cranes and quick hitch devices used on excavators as illustrated below.

204. When specific defects or unsafe circumstances are identified in fatal accident investigations, the need to alert the industry has to be considered alongside any 'live' legal case as action could impact on the case, or lead to HSE being sued. The HSE Board indicates that it takes a robust view that safety alerts take priority and is considering ways in which to make safety alerts more effective and disseminated more widely.

205. Plant and equipment are covered under interlaced European provisions requiring free movement of goods produced against harmonised standards. If a member state considers a harmonised standard is not sufficiently safe it can, and the UK does, raise the alarm at European level by means of a 'safeguard' action, with recent examples relating to fall arrest equipment and tower cranes.

Tower Cranes

206. Work on tower crane accidents has been pursued in two ways. Firstly, as this Inquiry was proceeding, the HSE Board agreed to recommend a package of measures including a voluntary register of tower cranes which would be made compulsory sometime in 2010. Secondly, the Strategic Forum's Tower Crane Working Group has produced guidelines for installation and operation to encourage best practice. As this work is developing separately from the Inquiry the issue of tower cranes, will not be covered in detail here.

Quick Hitch Devices used on Excavators

207. Semi-automatic quick hitches rely for their final security on an operator inserting a pin or similar locking or securing mechanism into the device itself. Failure to insert the pin has been implicated in a number of fatal accidents. HSE took action to highlight the problem to the industry[vv] and set out the steps needed to ensure that existing equipment could continue to be used in safety. In addition contractors, suppliers and manufacturers gave their commitment to supporting HSE in challenging the design of such equipment. The industry is undertaking, further work on quick hitches through the Strategic Forum's Plant Safety Group.

208. Although new semi-automatic quick hitch systems will not be supplied here in future, there are still thousands in use and available through hire and second-hand at auctions, some being supplied without pins. It will be years before they are phased out and, despite efforts to publicise, warn and educate, it is almost certain serious accidents and possibly fatalities will not be eliminated. Legacy issues present a significant challenge.

Future Improvement

209. The appropriate selection of equipment, taking account of the need to avoid, reduce and mitigate risk, is essential. In view of the commonality of serious and fatal accidents in certain areas more work should be done on:

   a) Suitable access equipment to enable fragile roof repairs to be undertaken from underneath.

   b) Improvement in systems and equipment to ensure the safe off-loading of delivery vehicles on construction sites.

[vv] Health and Safety Executive. 'HSE Safety Alert on quick hitches', December 2007. http://www.hse.gov.uk/construction/quickhitch.htm

c) Stairs rather than ladders to provide the normal means of accessing scaffolding wherever appropriate.

d) Developing alternatives to ladders such as more transportable, simple staging / scaffolding systems which can be used for domestic-type work and which provide guard-rails and platforms to prevent people falling.

e) Advances and increases in off-site construction / assembly, pre-fabrication.

f) Continued improvement in the design and use of scaffolding with advanced guard-rail systems.

g) Improvement in control systems on MEWPs to take account of basic human factors principles.

h) Improved detection of underground services.

i) Improvements in design, sequencing, fire engineering practices, packaging materials to improve fire precautions at <u>all</u> stages of build to protect the workforce.

WORKING WELL TOGETHER (WWT)

210. Working Well Together (WWT) is a good example of the industry and HSE working together to spread good practice particularly in the use of workplace equipment. It is celebrating its 10th Anniversary this year. It has 15 Regional groupings and the WWT Steering Committee is a sub-group of CONIAC, to which it reports. CONIAC is an advisory committee established by HSC to advise it in connection with regulatory policy and practice and other relevant matters. WWT Bus and roadshows have reached 41,000 people in its 10 years of existence and its practical exhibitions and seminars known as SHADs (Safety and Health Awareness Days) have seen about 25,000 people attending. It has its own website which acts as a portal to events and to useful material such as the Absolutely Essential Toolkit[ww].

211. As Philip White, the HSE's Chief Inspector of Construction, pointed out at the WWT's 10th Anniversary event, not everyone had a mobile phone in 1999 and Google was not yet one year old. Considerable changes have taken place which affect the industry and different priorities are adopted. In the immediate future WWT will be concentrating on Small Sites some of which do not accept they have duties to the workers on their sites. WWT has the advantage that it is not branded as part of 'officialdom' as the HSE is likely to be and would have more credibility with the small players in the industry.

212. WWT provides a good opportunity for the industry to promote partnership, working with companies through the supply chain. I believe that the industry should continue to support these developments as the more that is done, the more likely the industry generally will improve.

[ww] Health and Safety Executive. 'The absolutely essential health and safety toolkit for the smaller construction contractor', HSE Publication, INDG344 (rev2), July 2008. http://www.hse.gov.uk/pubns/indg344.pdf

HSE

213. The terms of reference of the Inquiry did not cover directly the work of the Health and Safety Executive or its Construction Division. Nevertheless its work is crucial to safety management and compliance, enforcement and, to an extent, culture change. It would have been inappropriate to conduct the kind of detailed study which has already been done by the Work and Pensions Committee[xx] or indeed by the HSE itself. Nevertheless, many stakeholders had views about its work and it is an extremely important resource for information and advice.

214. What is clear from the studies is that the HSE provides the most comprehensive information on safety in construction that we have. It might contain shortfalls, which the HSE readily accepts. However, it is the best we have. We know that under-reporting of even serious accidents is a problem in all areas and perhaps surprisingly that it is proportionately better in construction than the all industry average[yy]. It is a disgrace that we have such a low level of reporting of serious accidents, let alone near-misses, and is indicative, not just of ignorance and commercial factors for both the worker and the company, but of society's attitude to workers' injuries. If we had a higher proportion of reporting of serious accidents, it might help us to achieve a more accurate picture about fatalities. It would be ironic if the downward trend in fatalities is due entirely to the skills of the National Health Service in keeping people alive who might otherwise have died 25 years ago. Without better and more comprehensive reporting of serious accidents, irrespective of employment status, we cannot know. I recommend an awareness raising campaign so that individual workers and companies take the issue of reporting accidents more seriously.

215. In the absence of this picture the HSE has to work with the statistics it has, however inadequate. Even with fatalities it is quite shocking to learn that some only came to the attention of the HSE from the police or individual family members or a hospital (well after the event) and might otherwise not have been recorded as a construction fatality. Although the evidence is anecdotal, inspectors felt that some non-reporting was due to ignorance but in at least one case it was part of a cover-up.

216. The HSE's statistics base fatal accident rates to workers and injury rates to employees on figures from the Office of National Statistics (ONS). The Labour Force Survey confirms that under-reporting of serious injuries by the self-employed is much more significant than for

---

[xx] Department for Work and Pensions. 'The role of the Health and Safety Commission and Health and Safety Executive in regulating workplace health and safety', Third Report of Session 2007-08, HC 246, 2008

[yy] http://www.hse.gov.uk/statistics/pdf/ier.pdf

employees. For this reason although HSE presents fatal accident rates for workers, RIDDOR[zz] reported injury rates are based on figures for employees. If they were used to present a picture for all workers in construction it would give a comparatively low and misleading figure.

217. One important change of policy in the HSE has been in its engagement with the construction industry. This has been welcomed by the vast majority of stakeholders I have spoken to. It has led to a better understanding by the industry of the HSE's mission and responsibilities and it has assisted HSE's Construction Division to concentrate its resources and influencing skills in areas which are most effective. The HSE's role has always been advisor and enforcer to some extent but this engagement with industry has required a change in approach and means that resources are spread more thinly. The industry would welcome far more engagement but resources are such that the Construction Division has to try to balance the pro-active work with the re-active work i.e. investigations and prosecutions. HSE's objective is that the majority of time should be on pro-active work. In practice the re-active work has to take priority. Significant parts of the industry want to see more inspections and prosecutions by inspectors as they feel undermined by the poor practices of others.

218. The simple answer is more resources and more inspectors to do all this work. It is clear that the inspector resource had been allowed to slip below an acceptable level. This is being rectified with the possible exception of London. The question is what is an adequate resource and to what extent would it have an impact on fatalities? Statistics do not help us in this. This may be because there are too few inspectors. There appears to be no direct correlation between numbers of construction inspectors and the number of fatalities. There is no doubt that parts of the industry, trade unions and bereaved families would feel more reassured if there were more inspections and more prosecutions. The arguments are similar to those used about police on the beat. This is a matter of judgement. It is vitally important that the HSE's engagement policy is not promoted at the expense of its inspection and enforcement role. The question whether increasing the number of inspectors and inspections would, of itself, reduce fatal accidents is simply not known. There may well be other strong arguments for increasing the number of inspectors and some of these are outlined in Paragraph 227. That is a political decision for Government, and for HSE to deploy its resources accordingly.

219. An important change in the inspector's role is the time taken to prepare for investigations and prosecutions. The work is now more complex and

---

[zz] The Reporting of Injuries, Diseases and Dangerous Occurrences Regulations 1995, SI 1995/3163

drawn out compared with twenty years ago partly because the larger companies put a lot of money and effort into defending cases they are involved in for reputational reasons. Former inspectors have confirmed this. Now inspectors are often challenged at every stage, even when there is a guilty plea. The built-in delays of the legal system between police, coroner, Crown Prosecution Service, the HSE and the courts are now even more of a challenge and cases can take five years to come to a hearing. In one case they were on their fifth HSE Inspector because of the delays.

220. It was suggested to me by two stakeholders that inspectors should be stripped of their investigating and prosecuting role and should concentrate on inspections and issuing notices. My view is that this would be a seriously retrograde step. The inspector's presence on the site of the fatality ensures that there is some expertise to look for underlying causes as well as the 'smoking gun', and that potential witnesses are interviewed by someone with a working knowledge of site safety and procedures. The amount of time they spend on investigating and prosecuting does have an impact on the number of site inspections but it is too important a role to leave to others.

221. A case has been made to the Inquiry about allowing self-regulation for those companies with a reasonably good safety record. This in turn would allow HSE resources to be concentrated on the areas where there are consistently poor performers. While not wishing to rehearse all the arguments for and against, I am clear that this would be a retrograde step. While it is important to recognise that major contractors and some smaller companies take ownership of the risks in construction and lead culture change from the top and it is true that there are more fatal accidents on smaller sites, there are still accidents occurring on sites where major contractors have control. There is such a high degree of outsourcing of labour and professional skills and such a fragmented structure in the industry allowing for large variations of skills and experience in its workforce, that it would be an unacceptable risk to allow a greater degree of self-regulation. High risk areas should continue to be targeted by HSE as the best way to use its limited resources and to that extent the involvement with the 'safer' companies will be proportionate. When a system is as complex as that in the construction industry, the importance of external monitoring should not be underestimated.

222. An issue which was raised in some quarters was that inspectors vary in their judgements and their areas of particular interest which leads to confusion about priorities on site. I have spent a good deal of time with inspectors and principal inspectors both individually and collectively. I believe that first and foremost they are professionals. They work on

prescribed priorities outlined in the Construction Division's annual plan which might mean that sometimes there is a priority on welfare facilities, sometimes refurbishment, hitches on excavators, safe scaffolding etc. In other words, high risk factors are identified and prioritised to reflect national trends.

223. They also make judgement based on their knowledge of the site and the company and the risks involved. They may serve an improvement notice in one situation, a prohibition notice in another and a verbal warning or follow-up letter in another. Superficially these might appear to be identical situations. They are exercising their professional assessment about the likely continuation of risk, and the likelihood of compliance. To that extent their judgements may appear inconsistent. If an accident or fatality subsequently takes place that adds to the conviction that the inspector should have taken a firmer position on an earlier visit but that is the siren song of hindsight. There may also be some regional variation in practice but the construction industry also has regional variations. Some such as Scotland and the North East have a higher proportion of directly employed construction workers. Others, such as London and the South East, have a majority of self-employed both bogus and genuine, and a significant proportion of migrant labour. Some household names have a pattern of infringements across the regions which is only picked up by national intelligence and some well-known companies may have different practices on their sites in London and in other parts of the country for instance. Is it inconsistent to make different assessments or is it looking at each situation on its merits? I do not believe that it would be desirable for inspectors to be made into box-ticking wardens. Their independence and professionalism is important and should be retained.

224. The point was made by some stakeholders that HSE should prosecute more in areas where there has not been an accident but where unsafe practices have been identified. This would help to underpin the importance of prevention rather than investigation after the event. In informal discussion with inspectors it became clear that this was a question of priorities and proportionality. One made the point that if the award for a non-accident prosecution was a £500 fine, the amount of work involved would be disproportionate in terms of all his other responsibilities. It is possible that non-accident prosecutions can have an impact on behaviour irrespective of the fine and this may be worth exploring. This is an area where HSE management needs to give more clarity both to external stakeholders and inspectors. Similarly on the issue of fixed penalty fines for certain infringements there was a variety of views from favourable – it would be another weapon in the armoury particularly for small companies – to unfavourable – working out what

proof existed before imposing a fine might cause just as much work as a prosecution.

225. I believe that, while the questions of prosecution for non-accident infringements and fixed penalties might raise interesting points of discussion about priorities and the best use of resources, I am not convinced that it is a major factor in the underlying causes of fatalities and prefer to leave it to debate. I do believe however, that the value of a prosecution is that it can send very clear signals to the industry that it is not only accidents that can lead to prosecution. Pro-active prosecutions are not just about the potential level of the fine. The HSE might like to review its priorities to see if more capacity can be called upon to undertake pro-active prosecutions, possibly as a pilot study. However, I accept that this would require additional resources if it were to be undertaken.

226. Clearly the issue of resources is key and is the subject of continuous debate amongst stakeholders and in Government. HSE's most recent funding settlement provided for steady state, and resources in the Construction Division were maintained. The HSE has an enormous range of responsibilities in different sectors and it interacts with Government Departments in a number of ways. Although an important part of HSE's work, the Construction Division is one of many and HSE Inspectors are often re-deployed into different areas. That in itself leads to some frustration among inspectors who enjoy construction and prefer not to move. However, I believe it would be improper of me to try to direct the HSE's Board and Management about how it runs its organisation. As with any NDPB it is allocated resources and it takes decisions about annual priorities in discussion with Government. Demanding more money for construction might have the effect of depriving another part of the HSE of resources. As a regulatory agency, HSE requires sufficient resources to carry out its overall responsibilities and it is important that the HSE Board, and Government, give due weight to the demands made on HSE staff. HSE needs to deal effectively with the long-standing problem of inadequate numbers of inspectors in London. While there are a higher number of prestige sites in London there is also a higher number of what can only be described as cowboy operations.

227. Without wanting to make further specific recommendations on resources, I make the following points to try to be helpful:

a) Construction inspectors are an important resource and their numbers should be maintained at a level to ensure they carry out their current role effectively.

b) It should be recognised that considerably more of an inspector's time is taken on investigation and prosecution than in the past and there should be sufficient numbers to recognise that.

c) If Government and/or the HSE wish to make a concerted attempt at more non-accident prosecutions, it will require additional resource.

d) If there is to be more engagement with industry it should not be at the expense of site inspections and may require more resources.

e) If there is to be an increase in the proportion of serious accidents investigated by HSE, more resources will be required.

f) There is academic evidence that fatalities in construction rise as we come out of a recession. It is extremely important that sufficient resource is available to ensure that this is anticipated in the next few years by maintaining the cadre of inspectors.

228. I think there is an opportunity to review the communication of information about fatal accidents in a more co-ordinated and easily absorbable way. The information may be there on the web-site but I think presentation and targeting are very important. I think more opportunities could be found to publicise outcomes of legal cases and 'hot-spot' accidents. It may seem repetitive to the HSE but it will be new to someone in the industry.

229. I also recommend that the HSE continues the in-depth analysis of fatal accidents similar to the exercise carried out in Phase 2 of this Inquiry (see Paragraphs 231 to 238). The findings should be publicised to encourage both the industry and HSE Inspectors to think more about the underlying causes of fatal accidents and prevention strategies. Transparent, well-presented information about causal factors, including employment status, experience, client involvement and site supervision, will be invaluable to client, contractor and to the HSE itself.

230. The IoD/HSC guidelines should be used more vigorously by the HSE in the short-term. They represent an important step forward in providing a framework for directors' responsibilities. I am aware that work is being done to adapt them for small companies and that development is welcome. If the guidelines are to penetrate further into the industry, a clear communications strategy is required and inspectors should be supported with easily absorbable material to distribute.

CASE STUDIES

231. The Review and sample analysis of recent construction fatalities[9] is, in my view, an important piece of work. It represents a thorough and structured effort to clarify what the high level factors for construction fatalities might be. Before the exercise the peer reviewers and myself agreed on the factors which were important in determining the underlying causes of construction fatal accidents.

a) Economic indicators

b) Technology change

c) Rate per 100,000 workers

d) Structural changes in the industry e.g. increased use of sub-contracting, organisation of supply chains

e) Type of accident e.g. falls from heights

f) Hot-spots or success factors

g) Stage of the building project

h) Type of site (small / large, public / private)

i) Employment status / occupational mix, hours worked, experience, competency, time on site, familiarity of roles

j) Regulation, inspection, scrutiny

k) Complexity, management skills

l) Vulnerability.

232. Our objective was to identify the high risk categories. If this could not be done was it because of the absence of statistics or research, and, if so, what missing data was important which would allow such identification? Or was the problem universal in construction?

233. We were interested particularly in the following categories of relative risk exposure:

a) High fatality / low exposure

b) High fatality / high exposure

c) Low fatality / high exposure.

234. The approaches to tackling these underlying causes would be different and would identify possible solutions e.g. if low fatalities were due to good control. Do we know enough about the high categories to understand why the risks are not controlled? Do we understand what would control those risks? Do we understand the barriers such as the following to controlling risk?

a) Getting knowledge to the right people

b) Organisational problem of implementation

c) Incentives

d) Resources.

What changes are required to overcome those barriers?

235. As the Review makes clear[g], we examined 26 construction accidents leading to 28 deaths in the three year period 2005-06 to 2007-08 during which 211 men died in the industry in total. We emphasised that the sample is too small for statistically robust conclusions to be drawn and that the focus was on qualitative insights.

236. One area which was not clear in the Review was whether any of the deceased were members of trade unions or whether any of the employers (where they existed) were members of trade groups or federations. Although the factual information is not available, it is safe to assume that very few, if any, were trade union members or members of trade associations. However, in the absence of specific information, it would be unwise to draw any inferences from this.

237. The findings of the Review reveal a combination of factors for each fatal accident. However, the following are the most frequently cited:

a) The incidence of training factors, experience

b) Information and advice deficiencies

c) Risk perception

d) Rescheduling of work without planning

e) Minor / one-off jobs

f) Compliance

g) Equipment operability, space, PPE issues and tools not designed to fit the user / task.

238. What is clear is that most, if not all, of the accidents were preventable. The industry knows what it should be doing, the safety regulations and advice and guidance exist. This indicates to me that what is often missing is a safety culture, or adequate training and experience or supervision.

FURTHER RESEARCH

239. As the Inquiry progressed, the academic peer reviewers and I identified areas where it was not possible to reach definitive conclusions because of inadequate data or where there was insufficient coverage of a particular topic. In offering up the following suggestions for further work, we are not making claims that any or all of them would lead to a reduction in fatal accidents. However, we do believe the research topics might help to produce some practical solutions for the future.

Statistics

a) In depth analysis of a much larger sample of fatal accidents – perfecting the analysis tool used in Phase 2 case studies (retrospectively and prospectively) devising most effective feedback mechanisms to inspectors and the industry.

b) Research into robust measures of exposure to risk so that different influencing factors can be assessed on the basis of rate rather than absolute numbers e.g. economic activity in the industry and its different sectors, employment, size of company etc.

Hardware

c) Technological development in relation to the tools and equipment (and methods of work with them) particularly those most involved in serious accidents and occupational health risks. How can innovative design be encouraged?

d) Improvement in systems and equipment to ensure the safe off-loading of delivery vehicles on construction sites.

e) Developing alternatives to ladders such as more transportable simple staging / scaffolding systems which can be used for domestic type work and which provide guard-rails and platforms to prevent people falling.

f) Continued improvement in the design and use of scaffolding with advanced guard-rail systems.

g) Improvement in control systems on MEWPS to take account of basic human factors principles.

Training

h) Apprentices – The reasons for the high number of applications and high drop out rate.

International Comparisons

i) International good practice and correlation with fatal accident figures.

j) Devising comparable statistics on underlying causes.

Legal System

k) A study on the built-in delays in the system for handling fatal accident inquiries and prosecutions.

l) Influencing factors leading to prosecutions compared with those which do not.

m) Pilot scheme to increase number of pro-active prosecutions i.e. in cases where no accident has taken place.

n) Research to identify patterns of prosecution of Principal Contractors and / or sub-contractors in the event of a fatal accident (ref Paragraph 76). To identify whether the fragmented nature of the construction industry creates unintended outcomes in identifying responsibility for a fatal accident.

Groups

o) A study of sole traders along the lines of the small Loughborough University sample to understand what influences them and how they understand risk.

Reporting

p) Developing and promulgating an accident and incident reporting and analysis tool simple enough for medium sized firms.

q) Investigating the possibility of applying dangerous situation reporting in construction to encourage pro-active reporting.

Standards

r) Certification and standards and their use and misuse in construction. This covers people, equipment and organisations. What can be learned from the current successes and manifest failures? What can be done about this as a coherent tool for regulation?

s) Can safety be ring-fenced from financial production and deadline pressures? Examples of good practice on how it can work.

t) Paperless auditing of activities. Is it feasible and if so how?

u) Collection and analysis of good practice cases from construction and comparable industries with a major supply chain. Good practice in workplace engagement, flexible planning tools, worker consultation, workers safety representatives, professional safety advice, supply chain management.

Insurance

v) Research into how the insurance industry impacts upon safety in construction. How insurance companies measure safety and whether they adopt discounting for good safety on site, and whether sponsorship of the safety advisors would be of mutual benefit to the insurance industry and to the construction industry.

Vulnerable Workers

w) Deeper studies of groups of potentially vulnerable workers in the construction industry (e.g. migrant, young, old, uncertificated) and ways out of vulnerability.

x) The concept of 'vulnerability' requires more robust understanding – whether there are supporting pathways out of vulnerability or circumstances in the industry which act as traps in which vulnerability is increased.

Occupational Health

y) The dimensions of occupational health remain ill-defined. Further work is needed particularly for precarious workers. International comparisons for analogous workers (e.g. contract workers in nuclear and chemical industries) suggest extreme cases of vulnerability and a hidden iceberg of occupational mortality and morbidity.

Trade Unions

z) Challenges facing trade unions in UK construction – exploring how their important contribution as a resource to prevent accidents and ill-health at work could best be supported and utilised, particularly where traditional approaches to workplace representation might have relatively limited impact.

Crosscutting Research

240. There is a need for more crosscutting robust quantitative and qualitative research across a range of issues found in the industry. Most significantly, it is important that crosscutting study is undertaken to understand better how all these factors can be supported to operate most effectively in delivering the regulatory aims of CDM 2007 and those of EU requirements without the effects of bureaucratisation of risk management and audit that are currently observed. Crosscutting themes include:

a) How supply chains can work to the benefit or detriment of workers' health and safety.

b) What supports improved occupational health and safety practices in small firms and fragmented work situations.

c) How to achieve meaningful risk management in small firms.

d) A better understanding of the strengths and weaknesses of procurement strategies and certification systems, and the monitoring, auditing and 'enforcement' mechanisms for good practice in supply chains.

e) The role of intermediaries, actors and 'boundary spanning agents' in the organisational relations found in the industry.

APPENDIX 1 – INDEPENDENT ACADEMIC PEER REVIEWERS

Andrew Hale has been Professor of Safety Science at the Delft University of Technology in the Netherlands, full time from 1984-2007 and part-time thereafter. He is currently chairman of Hastam Ltd in UK. He has worked in the area of safety and health since 1966, initially on accident investigation and human behaviour in occupational safety, and later on professionalisation of the field of safety, on safety management and regulation, not only in occupational, but also transport safety. He has broad experience as a member of policy, advisory and evaluation committees for safety research institutions in several European countries and as chairman and member of a number of government advisory committees in the Netherlands. He was chief editor of Safety Science from 1993 until the end of 2008 and has been on the board of some half dozen other scientific journals.

Sonia McKay currently holds the post of Reader in Socio-Legal Studies at the Working Lives Research Institute where she heads a number of major research projects, mainly focusing on migration. She was the principal author of research report: *Migrant workers in England and Wales – an assessment of migrant worker health and safety risks*. She also leads the EU Level Team for the Observatories at the European Foundation for Living and Working Conditions. She holds a law degree from Queens University, Belfast and a Ph.D in employment law from Wolfson College, Cambridge. Sonia came to the Working Lives Research Institute from the Labour Research Department (LRD), the independent trade union-based research organisation, where she held the post of employment law researcher from 1983.

David Walters is Professor of Work Environment, Director of Cardiff Work Environment Research Centre (CWERC), Cardiff School of Social Sciences, Cardiff University. His research and writing relate to various aspects of the work environment with particular interests in employee representation and consultation on health and safety, the politics of health and safety at work, regulating health and safety management, chemical risk management and health at work, safety in small firms and maritime health and safety management. Recent research includes a study of the role of supply chains in regulating health and safety, knowledge transfer in chemical risk management in small firms, the impact of the EU Chemical Agents Directive in EU member states and ongoing international studies on worker representation and the working environment. He is founding editor of the journal 'Policy and Practice in Health and Safety'.

APPENDIX 2 – CONSULTATION.

Organisations[aaa]

Association of British Insurers (ABI)

Stephen Hadrill	Director General
Matthew Young	Policy Advisor – Liability and Occupational Health

Association for Project Safety (APS)

Philip Baker	President
Brian Law	Chief Executive
Stella Saunders	Chair – Wales Region

BAA

Mike Evans	Head of Health and Safety Capital Projects

BAM Construct UK Ltd

Frank Garnett	Director of Health and Safety

British Safety Council (BSC)

Brian Nimick	Chief Executive
Neal Stone	Head of Policy and Public Affairs
Shaun Davis	Group Director for Health, Safety and Environment – Rok plc
John Morgan	Director for Safety, Health and Environment – Kier Group plc
Donna Newell	Director for Health and Safety and Human Resources – Rydon Group Ltd

Civil Engineering Contractors Association (CECA) Wales

Huw Llywelyn	Chair Chair – South Wales WWT Managing Director – Alun Griffiths (Contractors) Ltd

Centre for Corporate Accountability (CCA)

David Bergman	Director

Confederation of British Industry (CBI)

Neil Carberry	Head of Health and Safety Policy
Richard Cummings	Policy Advisor

Constructing Excellence (CE) in the Built Environment

Tom Harper MBE	Director – South West
Don Ward	Chief Executive

Constructing Excellence in Wales (CEW)

Milica Kitson	Chief Executive

[aaa] This appendix lists the people the Chair spoke to on the telephone or in meetings during the Inquiry – some represented multiple organisations and where relevant and as far as possible this is reflected in the presentation.

Construction Clients' Group (CCG) – Constructing Excellence

James Preston-Hood	Chair of Health and Safety Sub-Group Health and Safety Director - Grosvenor

Construction Industry Advisory Committee (CONIAC)

Meeting attended on 26 March 2009

Construction Industry Council (CIC)

Peter Caplehorn	Health and Safety Committee Technical Director – Scott Brownrigg Ltd
Gordon Masterton OBE	Deputy Chairman
Graham Watts OBE	Chief Executive

Construction Industry Training Board (CITB)

Sir Michael Latham	Chairman

ConstructionSkills – Formerly known as CITB

Kevin Fear	Head of Health, Safety and Environmental

Construction Safety Campaign

Tony O'Brien	Secretary
Peter Farrell	Chair

Construction Skills Certification Scheme (CSCS)

Trevor Walker	Chairman Chairman – Constructing Better Health (CBH)

Coroners' Society of England and Wales

Caroline Beasley-Murray	HM Coroner for Essex and Thurrock

Department for Business, Innovation and Skills (BIS) - formerly BERR

Rt Hon Patrick McFadden MP	Minister for Employment Relations and Postal Affairs (until 8 June 2009; currently Minister for Business, Innovation and Skills)
Geoffrey Norris	Special Advisor to Secretary of State
Ian Pearson MP	Economic and Business Minister (until 8 June 2009)
Denis Walker	Director for Construction Sector Unit
Clive Young	Assistant Director in the Construction Sector Unit

Department for Communities and Local Government (CLG)

Anthony Burd	Head of Profession
Sarah Sturrock	Head of Sustainable Buildings Division

Department of Health (DH)

Ray Stephenson	ProCure 21 Programme Manager

Department for Work and Pensions (DWP)

Rt Hon Yvette Cooper MP	Secretary of State (from 9 June 2009)
Lord McKenzie of Luton	Parliamentary under Secretary of State (Lords)
Sir Leigh Lewis	Permanent Secretary
Rt Hon James Purnell MP	Secretary of State (until 4 June 2009)
Teresa Quinn	Health and Safety Sponsorship Team (from 2 March 2009)
Philip White	Health and Safety Sponsorship Team (until 15 March 2009)

Engineering Construction Industry Association (ECIA)

Richard Ash	Safety, Health and Environmental Manager

Employment Agency Standards Inspectorate (EASI) – part of BIS

Carl Cresswell	Assistant Director of Guidance and Enforcement
Steven Keeler	Investigation Manager

Federation of Master Builders (FMB)

Brian Berry	Director of External Affairs
Richard Diment	Director General
Peter O'Connell	Policy Manager

Gangmasters Licensing Authority (GLA)

Paul Whitehouse	Chairman

GMB

John McClean	National Health and Safety Officer

Health and Safety Executive (HSE)

David Ashton	Director of Field Operations Division
Roxanne Barker	HM Inspector of Health and Safety
Sally Bates	Compliance Officer
Ian Betley	HM Inspector of Health and Safety
Brent Bolton	HM Inspector of Health and Safety
Louise Brearey	Head of Construction Sector
Alex Brett-Holt	Legal Advisor
Stuart Bristow	International Unit
Richard Boland	Head of Operations - Construction Division
Lisa Chappell	HM Inspector of Health and Safety
Mike Cross	Head of Operations - Construction Division
Emma Davies	Construction Sector
Andrew East	HM Principal Specialist Inspector
Joanne Eccles	HM Inspector of Health and Safety
Rosi Edwards	Head of Operations - Construction Division
Mike Ford	HM Inspector of Health and Safety
Wendy Garnett	HM Inspector of Health and Safety
Judith Hackitt CBE	Chair
Mike Harrison	Construction Sector
Simon Hester	HM Inspector of Health and Safety
Elizabeth Hodkinson	Head of Strategic Interventions Division
Debbie John	HM Inspector of Health and Safety
Frances Kennedy	Health and Safety Awareness Officer
Andrew Kingscott	HM Principal Inspector of Health and Safety
Andrew Lambert	Health and Safety Awareness Officer
Anthony Lees	Construction Policy
Norman Macritchie	HM Inspector of Health and Safety
Yvonne Mazzotta	Health and Safety Awareness Officer
John Moutrie	HM Specialist Inspector
Kevin Myers	Deputy Chief Executive
Chantal Nicholls	HM Principal Inspector of Health and Safety
Sue Parkyn	Construction Sector
Geoffrey Podger CB	Chief Executive
Jeanette Reuben	Head of Operations - Construction Division
Tim Shambrook	Working Well Together Campaign Manager
Ian Simpson	HM Specialist Inspector
Jim Skilling	HM Principal Inspector of Health and Safety
Neil Stephens	Construction Sector
David Stewart	HM Inspector of Health and Safety
Paul Stollard	Director of Scotland
Pauline Storey	Compliance Officer
Steve Tampling	Construction Sector

Adrian Tinson	HM Inspector of Health and Safety
Bob Tunnicliffe	Construction Sector
Philip White	Chief Inspector of Construction (from 16 March 2009)
Stephen Williams	Chief Inspector of Construction (until 15 March 2009)
Dave Wonford	HM Specialist Inspector
Stephen Wright	Statistician

Health and Safety Executive Board

Meetings attended on 25 February and 27 May 2009

Health and Safety Executive for Northern Ireland (HSENI)

Kevin Toner	Deputy Chief Executive

Institution of Occupational Safety and Health (IOSH) – Construction Group

Richard Jones	Policy and Technical Director
John Lacey	Chair of Construction Group

Local Authority Building Control (LABC)

Paul Everall CBE	Chief Executive

Local Authorities Coordinators of Regulatory Services (LACoRS)

Mark Du Val	Director of Policy

National Assembly for Wales

Vincent Doyle	Senior Health and Safety Manager
Nigel Elias	Head of Capital Projects
Richard Wilson OBE	Construction Programme Director Chairman – Constructing Excellence in Wales

National Federation of Builders (NFB)

Paul Bogle	Policy Manager
Julia Evans	Chief Executive
Alan Muddiman	Health and Safety Director - Renew Holdings plc
Bill Rabbetts	Chairman & Managing Director – Herbert H Drew and Son Ltd

National Federation of Roofing Contractors (NFRC)

Andrew Clarke	Political Advisor – Ranelagh International Ltd
Ray Horwood CBE	Chief Executive
Mike Long	President and Chair of Health & Safety Committee

Office of Government Commerce (OGC)

John Ioannou	Head of Commercial Delivery, Construction Markets and Collaborative Procurement

Olympic Delivery Authority (ODA)

Lawrence Waterman	Head of Health and Safety

Public Sector Construction Clients Forum (PSCCF)

Dame Mavis McDonald	Chair

Quarry Products Association (QPA)

John Sheridan	Regional Director – QPA Scotland Chair – Scottish Strategic Forum Health and Safety Group

Scottish Trades Union Congress (STUC)

Ian Tasker	Assistant Secretary

Scottish Building Federation (SBF)

Michael Levack	Chief Executive

Sentencing Guidelines Secretariat

Kevin McCormac	Head of Secretariat

Site Safe Scotland

Grahame Barn	Director - Federation of Master Builders Scotland Scottish Strategic Forum
John Forster	Managing Director – Forster Roofing Services Ltd Scottish Homebuilders Forum
Brian Hume	Health, Safety and Environmental Manager – Balfour Beatty Regional Civil Engineering Ltd

Specialist Engineering Contractors' Group (SEC)

Bill Belshaw	Chairman of SEC Group Health and Safety Task Group
Professor Rudi Klein	Chief Executive
John Nelson	Executive Secretary

Strategic Forum for Construction (SFfC) Health and Safety Task Group

Meeting attended on 22 January 2009

Additional meetings held with:

John Spanswick CBE	Chairman Chairman – Bovis Lend Lease Group Non-Executive Board member - HSE

Trades Union Congress (TUC)

Brendan Barber	General Secretary
Hugh Robertson	Senior Policy Officer Non-Executive Board member - HSE

UK Contractors Group (UKCG)

Health and Safety Sub Group Meeting attended on 20 January 2009

Stephen Ratcliffe	Secretary

Additional meeting held with:

James Wates	Chairman Deputy Chairman – Wates Group Ltd

Union of Construction, Allied Trades and Technicians (UCATT)

Harry Frew	Regional Secretary - Scotland
Dave Gunter	Regional Organiser
Jim Kennedy	National Political Officer
Alan Ritchie	General Secretary

UNISON

Hope Daley	Head of Health and Safety

UNITE

Bob Blackman	National Secretary for Construction
Rob Miguel	Health, Safety and Education Officer

Value Wales – part of the National Assembly for Wales

Martin Sykes	Chief Executive Officer

Welbro Project Management Ltd

Graham Alexander	Director

Welsh Built Environment Forum

Chris Hughes	Executive Director

Work and Pensions Committee

Terry Rooney MP	Chairman

Inquiry Chair appeared before the Committee on 26 January 2009

Bereaved families and campaign groups

Battersea Crane Disaster Action Group (BCDAG)

Liliana Alexa	
Matthew Willis	Solicitor
Audrey Winter	

Families Against Corporate Killers (FACK)

Hilda Palmer	Facilitator
Linda Whelan	

Hazards Campaign

Judith Allen

Jennifer Deeney

Anthea Dennis	
Peter Dennis	
Mick Antinow (with Mr & Mrs Dennis)	Thompsons Solicitors

Individuals meet within the context of other events

Working Well Together event 17 February 2009

Building Safety Group Ltd

Brian Roebuck	Principal Safety Advisor

Denholm Industrial Services Ltd

Richard Noakes	Quality, Health and Technical Advisor
Damian Barlow	Contracts Manager

London Borough of Merton

Matthew Southgate	CHAS Scheme Business Development Manager

National Access and Scaffolding Confederation (NASC)

David Chapman	Health, Safety and Technical Advisor

National Construction College (NCE) East

David Gaughran	Manager - Access Department

UKCG Health and Safety Sub Group meeting 20 January 2009

Vaughan Burnand	Chairman – Major Contractors' Group Health and Safety Sub Group (until January 2009)

Academics

University of Westminster

Professor Linda Clarke	European Industrial Relations – Westminster Business School

University of Essex

Professor Mark Harvey	Director - Centre for Research in Economic Sociology and Innovation

University of Warwick

Professor Frank Wright	Warwick Law School

Legal Profession

Madeleine Abas	Senior partner - Osborn Abas Hunt
Richard Matthews	Chambers of William Clegg QC

Opposition Spokespersons

Rt Hon Teresa May MP	Shadow Secretary of State for Work & Pensions and Shadow Minister for Women & Equality Conservative
Paul Rowen MP	Shadow Work & Pensions Team Liberal Democrat

APPENDIX 3 – WRITTEN EVIDENCE

1.	1 February 2009	Mr Michael J Ponsonby (individual)
2.	13 February 2009	Specialist Engineering Contractors Group (SEC) Association of Plumbing & Heating Contractors British Constructional Steelwork Association Electrical Contractors Association Heating & Ventilating Contractors Association Lift and Escalator Industry Association SELECT (Electrical Contractors Association of Scotland)
3.	19 February 2009	UK Contractors Group (UKCG)
4.	9 March 2009	Mr Stephen Parris (individual)
5.	11 March 2009	Safety Pass Alliance (SPA) Ltd
6.	13 March 2009	British Safety Council (BSC)
7.	March 2009	Union of Construction, Allied Trades and Technicians (UCATT)
8.	18 March 2009	Union of Construction, Allied Trades and Technicians (UCATT)
9.	27 March 2009	National Federation of Builders (NFB)
10.	1 April 2009	Confederation of British Industry (CBI)
11.	21 May 2009	Health and Safety Executive (HSE)
12.	19 June 2009	National Federation of Roofing Contractors (NFRC)

APPENDIX 4 – BACKGROUND LITERATURE

Non-HSE (in date order)

1. Muddiman, A. 'The Construction Industry's response to deaths at work – Grasping the nettle'. MA dissertation. 2001.

2. Union of Construction Allied Trades and Technicians (UCATT). 'Worker Engagement in the Construction Industry'. October 2006.

3. Office of Government Commerce (OGC). 'Achieving Excellence in Construction Procurement Guide 01. Initiative into action'. 2007.

4. Office of Government Commerce (OGC). 'Achieving Excellence in Construction Procurement Guide 02. Project organisation: role and responsibilities'. 2007.

5. Office of Government Commerce (OGC). 'Achieving Excellence in Construction Procurement Guide 03. Project procurement lifecycle: the integrated process'. 2007.

6. Office of Government Commerce (OGC). 'Achieving Excellence in Construction Procurement Guide 04. Risk and value management'. 2007.

7. Office of Government Commerce (OGC). 'Achieving Excellence in Construction Procurement Guide 05. The integrated project team: team working and partnership'. 2007.

8. Office of Government Commerce (OGC). 'Achieving Excellence in Construction Procurement Guide 06. Procurement and contract strategies'. 2007.

9. Office of Government Commerce (OGC). 'Achieving Excellence in Construction Procurement Guide 07. Whole-life costing and cost management'. 2007.

10. Office of Government Commerce (OGC). 'Achieving Excellence in Construction Procurement Guide 08. Improving performance project evaluation and benchmarking'. 2007.

11. Office of Government Commerce (OGC). 'Achieving Excellence in Construction Procurement Guide 09. Design quality'. 2007.

12. Office of Government Commerce (OGC). 'Achieving Excellence in Construction Procurement Guide 10. Health and safety'. 2007.

13. Office of Government Commerce (OGC). 'Achieving Excellence in Construction Procurement Guide 11. Sustainability'. 2007.

14. European Institute for Construction Labour Research (CLR) News. Poland's Construction Industry. No 2/2007.

15. Centre for Corporate Accountability for UCATT. 'Levels of Convictions and sentencing following prosecutions arising from deaths of workers and members of the public in the Construction sector'. April 2007.

16. Trade Union Congress (TUC). 'Safety & Migrant Workers: A practical guide for safety representatives'. June 2007.

17. Department of Work and Pensions (In conjunction with HSE and ConstructionSkills). 'Achieving Behavioural Change Launch Event'. 25 July 2007.

18. Centre for Business Research - University of Cambridge. Simon Deakin & Aristea Koukiadaki. 'The Capability Approach and the Reception of European Social Policy in the UK: The Case of the Information and Consultation of Employees Directive – Section 5.4 Establishment and operation of information and consultation of employees arrangements: Heathrow Terminal 5'. September 2007.

19. Institute of Directors (IoD) and Health and Safety Commission (HSC). 'Leading health and safety at work: Leadership Actions for Directors and Board Members'. INDG417. October 2007.

20. Union of Construction Allied Trades and Technicians (UCATT). 'Bringing justice to the Boardroom – The case against voluntary guidance and in favour of a change in the law to Improve Safety Duties on Directors'. October 2007.

21. National Audit Office (NAO). 'Department of Work and Pensions: The roll-out of the Jobcentre Plus Office network'. HC 346 Session 2007-2008. 22 February 2008.

22. House of Commons Work and Pensions Committee. 'The Role of the Health and Safety Commission and the Health and Safety Executive in workplace health and safety'. Third report of Session 2007-2008: Volume II. Oral and written evidence. 21 April 2008.

23. Union of Construction Allied Trades and Technicians (UCATT). Mark Harvey & Felix Behling. 'The Evasion Economy: False Self-Employment in the UK Construction Industry'. 14 May 2008.

24. Westminster Business School, University of Westminster. Linda Clarke and Maria Gribling. 'Obstacles to Diversity in Construction: the example of Heathrow Terminal 5'. Published in Construction Management and Economics. July 2008.

25. House of Commons Work and Pensions Committee. 'The role of Health and Safety Commission and the Health and Safety Executive in workplace health and safety: Government Response to the Committee's Third Report of Session 2007/2008'. HC 837. 2 July 2008.

26. Strategic Forum for Construction Health and Safety Task Group. 'Health and Safety Performance in the Construction Industry - Second Industry Report to HMG Response to the Construction Forum: Framework for Action'. August 2008.

27. Department for Business Enterprise & Regulatory Reform (BERR). 'Vulnerable Worker Enforcement Forum. Final Report and Government Conclusions'. August 2008.

28. National Access & Scaffolding Confederation (NASC). 'Interim Guidance on Collective Fall Prevention Systems in Scaffolding'. SG4: 05 Appendix A. September 2008.

29. The Association for Project Safety (APS). Practice Register Standard. 2008.

30. Department of Work and Pension (DWP). Transcript of DWP Select Committee - 'Evidence on the work of DWP's Inquiry into the Underlining Causes of Construction Fatalities with Rita Donaghy (Chair of Inquiry)'. January 2009.

31. Cardiff Work Environment Research Centre (CWERC). 'Paradoxes of the supply chain and worker's health and safety: An international seminar on the role of the supply chain in health and safety management and performance'. 3 March 2009. http://www.cardiff.ac.uk/cwerc/events/March%20seminar/supply-chain-workshop.html

32. Department of Health website. NHS Overview of Procurement 21. www.nhs-procure21.gov.uk/content/overview.asp. March 2009.

33. Irwin Mitchell Solicitors and the Centre for Corporate Accountability (Commissioned and Jointly Published the report). 'Migrant Workplace Deaths in Britain'. 31 March 2009.

34. Scottish TUC. Improving Occupational Health in Scotland.

HSE (in date order)

1. HSE. 'One Hundred Fatal Accidents in Construction'. A Report by HM Factory Inspectorate. 1978.

2. HSE. 'Blackspot Construction'. A Study of Five Years Fatal Accidents in the Building and Civil Engineering Industries'. 1988.

3. HSE. 'Reducing error and influencing behaviour'. HSG48. 2nd Edition, June 1999.

4. HSE. 'Causal Factors in construction accidents'. Department of Human sciences and Civil and Building Engineering Loughborough University, Manchester Centre for Civil and Constriction Engineering UMIST for the HSE. RR156. 2003.

5. Work-related deaths: A protocol for liaison. MISC491. 2nd Edition, February 2003.

6. HSE. 'The High 5 – Five ways to reduce risk on site for small construction sites / business / workers'. Jointly with WWT. INDG384. September 2003.

7. HSE. 'A short guide to Managing Asbestos in premises'. INDG223 (rev3). 2002, reprinted 2004.

8. HSE. 'Enforcement Management Model'. Operational version 3.0. September 2005.

9. HSE. 'Top tips for ladder and stepladder safety'. INDG405. October 2005.

10. HSE. 'Work-related deaths: A protocol for liaison - Scotland'. MISC733. September 2006.

11. HSE. 'Health and safety in construction'. HSG150. 3rd Edition, April 2006.

12. HSE. 'Essentials – of health and safety at work'. 4th Edition, April 2006.

13. HSE. 'Five steps to risk assessment'. INDG163 (rev2). June 2006.

14. HSE. 'The Work at Height Regulations 2005 (as amended)'. INDG401 (rev1). April 2007.

15. HSE. 'Getting to grips with Manual Handling – A short guide'. INDG143 (rev2). 2004, reprinted September 2007.

16. HSC. 'Managing health and safety in construction: Construction (Design and Management) Regulations 2007, Approved Code of Practice. L144. November 2007.

17. HSE. 'Asbestos Kills: Protect yourself! You are more at risk than you think'. INDG419. December 2007.

18. HSE. 'Management of Health and Safety at Work Regulations 1999'. Approved Code of Practice and Guidance. L21. 2nd Edition 2000, reprinted December 2007.

19. HSE. 'Construction Intelligence Report: Analysis of Construction Injury and Ill Health Intelligence'. 1997/98 - 2007/08.

20. HSE. 'Asbestos Essentials – A task manual for building , maintenance and allied trades on non licensed asbestos work'. HSG210. 2nd Edition, February 2008.

21. HSE. 'The Absolutely essential health and safety toolkit – for the smaller construction contractor'. INDG344 (rev2). Jointly with WWT. July 2008.

22. HSE. 'An Introduction to Health and Safety – Health and safety in small businesses: what you should know about – where to get more information'. INDG259 (rev1). 2003, reprinted August 2008.

23. HSE. 'Successful health and safety management'. HSG65. 1997, reprinted with amendments 2000, reprinted December 2008.

24. HSE. 'Health and safety in roof work'. HSG33. 3rd Edition, December 2008.

25. HSE. 'Scaffold Checklist'. www.hse.gov.uk/construction/scaffoldinginfo.htm. 23 January 2009.

26. HSE. 'Qualitative Research amongst 'Hard to Reach' Small Construction Site Operators for the Health and Safety Executive'. Corr Willbourn Research & Development. 19 February 2009. To be published.

27. HSE. 'Enforcement policy statement'. HSE 41 (rev1). February 2009.

28. HSE. 'The safe use of vehicles on construction sites – a guide for clients, designers, contractors, managers and workers involved with construction transport'. HSG144. 2nd Edition, April 2009.

29. HSE. 'Construction Division Plan of Work 2009-10'. 2009.

30. Health and Safety Executive website. Operational procedures: Investigation. http://www.hse.gov.uk/foi/internalops/og/ogprocedures/investigation/

31. HSE. 'The Health and Safety of Great Britain \\ Be part of the solution'. June 2009.

APPENDIX 5 – KEY REGULATIONS IN CONSTRUCTION

1. The Safety Representatives and Safety Committees Regulations 1977, SI 1977/500.

2. The Electricity at Work Regulations 1989 (as amended), SI 1989/635.

3. The Personal Protective Equipment at Work Regulations 1992 (as amended), SI 1992/2966.

4. The Manual Handling Operations Regulations 1992 (as amended), SI 1992/2793.

5. The Workplace (Health, Safety and Welfare) Regulations 1992 (as amended), SI 1992/3004.

6. The Reporting of Injuries, Diseases and Dangerous Occurrences Regulations 1995, SI 1995/3163.

7. The Health and Safety (Consultation with Employees) Regulations 1996, SI 1996/1513.

8. The Provision and Use of Work Equipment Regulations 1998 (as amended), SI 1998/2306.

9. The Lifting Operations and Lifting Equipment Regulations 1998 (as amended), SI 1998/2307.

10. The Health and Safety (Enforcing Authority) Regulations 1998, SI 1998/494.

11. The Management of Health and Safety at Work Regulations 1999 (as amended) SI 1999/3242.

12. The Control of Substances Hazardous to Health Regulations 2002 (as amended), SI 2002/2677.

13. The Information and Consultation of Employees Regulations 2004, SI 2004/3426.

14. The Work at Height Regulations 2005 (as amended), SI 2005/735.

15. The Control of Noise at Work Regulations 2005, SI 2005/1643.

16. The Control of Asbestos Regulations 2006, SI 2006/2739.

17. The Construction (Design and Management) Regulations 2007, SI 2007/320.

This list is not exhaustive and should not be used as a definitive list of health and safety regulations applicable to construction work.

Printed in the UK by The Stationery Office Limited
on behalf of the Controller of Her Majesty's Stationery Office
ID6178778 07/09 431833 19585

Printed on Paper containing 75% recycled fibre content minimum.